contemporary
landscape architecture

BRAUN

chris van uffelen (ed.)

contemporary
landscape architecture
masterpieces around the world

BRAUN

CONTENT

CONTENT

Americas

REGIONS

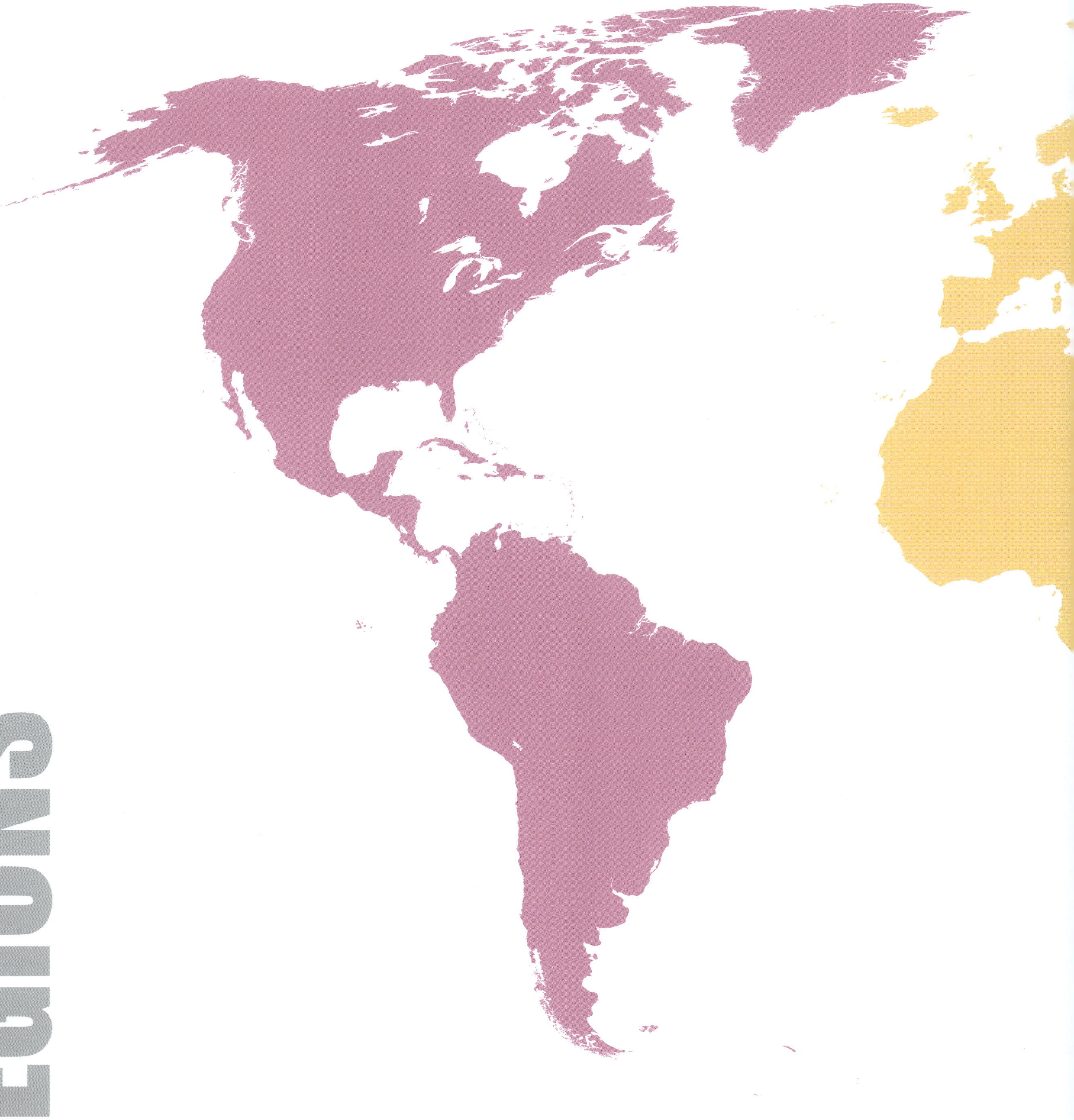

PREFACE

Landscape architecture and open space design have developed into an enormously important field of activity in recent decades. Less than half a century ago, with the exception of public parks and private gardens, they were often of purely secondary nature that served the built environment. Signs saying "Don't step on the grass" were a symbol of this in many public parks and open spaces in the last quarter of the 20th century. Unused lawns usually had the function of keeping the space in front of buildings free – a short-cropped green carpet, bordered or centered with a few flowers. Of course, this form of urban greening is still justified, but it also has new functions to fulfill.

For example, ecological functions. Considering the climate crisis, every form of open space design must also face the question of how it contributes to overcoming it. For example, the lawn on a palace square can also serve as an infiltration area or as physical tank for rainwater. Even school playgrounds can serve as a defense against flooding with sophisticated water retention systems without neglecting their actual task of giving students a break. Green open spaces must also contribute to biodiversity. Today, biodiversity in cities is often higher than in their surrounding areas. But this can certainly be further increased in the future with the right planting. Short-cropped lawns are not conducive to this. Nevertheless, they shouldn't be dismissed as "green concrete": On the one hand they are still greener than concrete itself, and on the other hand the visually calm open spaces are also important for the perception of space – not only in the context of a listed English Garden.

Many social tasks of open space design have also been added since the invention of the pedestrian zone. This includes, if not primarily, traffic calming, but centers on the creation of spaces of connection. Even before that, movement in spaces was directed by means of design – for example by the nature or design of the pavement – but since the pedestrian zone, transit was no longer the sole purpose. Every open space had to become a place to spend time, which was previously reserved for plazas, parks or gardens. The public demands these qualities, even in places where other functions are the reason for the design. Since 9/11/2001, one of these additional functions that the design of open space must provide is, for example, the guarantee of security against attacks.

Today, the profession is an extremely responsible one, which has to deal with numerous problems of our time. This volume presents around 300 answers from around the world. It also shows that the challenges are similar around the world, even if the characteristics may differ. The answers to those challenges are of very different kinds with differentiated approaches The volume shows the wide range of tasks as well. The abundance shows almost the entire spectrum with which landscape architecture offers solutions to the questions of our time: from small private gardens to company campuses, from village squares to city centers, from stream banks to sponge cities.

Quinta do Lago Villa

Wilderness in the Portuguese sense, regiao selvagem, or the Mediterranean scrubland areas of scrub and gravel, is the design concept of the entire garden. It is dedicated to a return to a wilder gardening experience, using a range of native or native-like species to create a more sustainable landscape with a diversity of experience. Sharply defined terraces, decks, and screening walls contrast with the planted landscape to create a play of light that changes throughout the day. The textured foliage plantings include grasses, creeping groundcovers, herbs, gnarled shrubs and trees, and tall pines. Boulders and stone walls are used to maintain grade and reflect the local vernacular. They also provide a counterpoint to the loose and naturalistic planting, particularly in the "dry stream" paths that run through the garden.

Villa Al-Thea

A modern finca with beautiful garden rises in the enchanting old cultural landscape of southern Spain with its olive trees, orange groves, and agricultural terraces. Designed by Studio Nico Wissing, the stacked drywalls, bancales de piedra, form the basis of the interplay of lines in the idyllic garden measuring 20,000 square meters. The smooth transition between the finca and its environment is the strength of the design. A long passerelle connects the house with the surrounding flower fields. The garden fits into the environment by only using native trees and shrubs from the region. Almost 60 percent of the area is sown with wild flowers and herbs. Thousands of flowers dancing in the wind, countless butterflies, insects, and birds make it a vibrant scenery.

Can Climent Square

The project was born as an intermediate step between a motorcycle parking lot and a forest in the heart of Barcelona. Planned by Sergi Carulla and Oscar Blasco from scob, the focus lay on converting the 2,800-square-meter area from a place of passage, without structure or human scale, into a space that can accommodate neighborhood life. Through provisional urban elements, the new Can Climent Square is delimited and adapted as a new meeting point, place of rest, or shared playing field for sporting activities. The color orange was used as a symbol of energy, optimism, and good humor. The aim of the project, commissioned by the Barcelona City Council, is to give the local neighborhood the enthusiasm to reclaim the Can Climent Square and reshape it by gathering as a community.

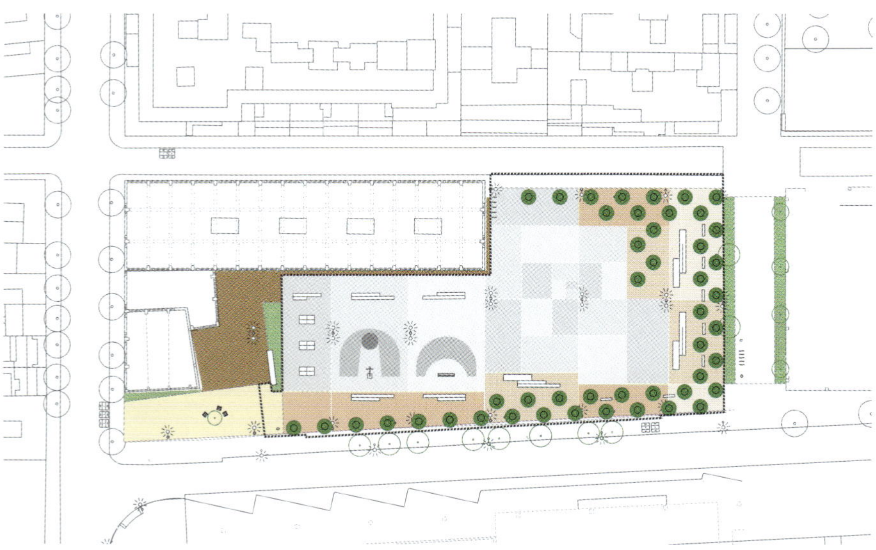

Can Bada Park

Can Bada Park is part of the Metropolitan Parks Network that manages the Barcelona Metropolitan area, together with the Torrent de la Font and Turó de l'Enric. To minimize earth movements, the park project uses terraces giving each one of them a specific use. An important point of the project is the collection of rainwater from the surface of the park for its irrigation. The vegetation has been arranged into groups of plants with the same water requirements, so that irrigation can be programmed according to the needs of each group and thus water consumption can be adjusted. With its minimal use of paved surfaces, the park positively counteracts all the other ones around the neighborhood, becoming a new green lung for Badalona. The plants in the park are all native species adapted to the climatic conditions of Badalona.

TB House

TB House, an architecture project by Rafael Beneytez of z4z4, is located at the edge of an urbanization in Avaraca, very close to the Pardo Mountain. El Pardo is a natural area and constitutes an ecological and biological reserve as a typical Mediterranean holm oak forest. The garden surrounding TB House has a wild appearance that blends in perfectly with the landscape. Granite boulders have been placed between a series of plots that mix perennials and grasses. An earthen path to the house and small pedestrian paths are embedded in the landscape. More than 50 varieties of plants were used to add color, texture, and movement to the garden. The garden extends to the inside of the house in the gap between two large walls, changing appearance and providing living areas for the family's daily life.

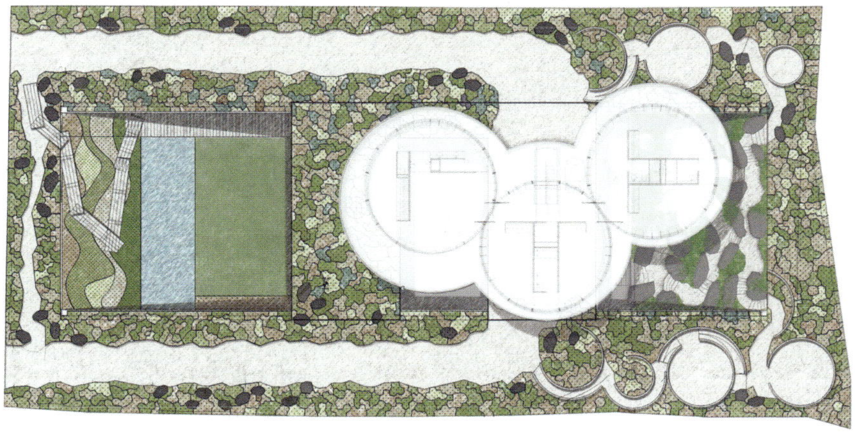

Coastal Walk in Costa Brava

Spains Costa Brava coast faces the double pressure of climate change and massive frequentation. The project aimed to connect and restore four kilometers of coastal walk landscape mosaic for the coming decades. Currently, the coastal path connects the center of Palamós with Cap Roig Natural Park and a landscape seasonal dissuasive parking. The design defines 38 specific cross-sections, 18 viewpoints, and up to six formulations of paving solutions working with different formulations of sauló, the local orange granite rock. Thus, the landscape is sublimated through minimal ad hoc interventions that highlight the peculiar site's features in a sequential path with permanent views of the sea. Along the path, degraded habitats with invasive flora have been restored to a finer and richer coastal landscape mosaic.

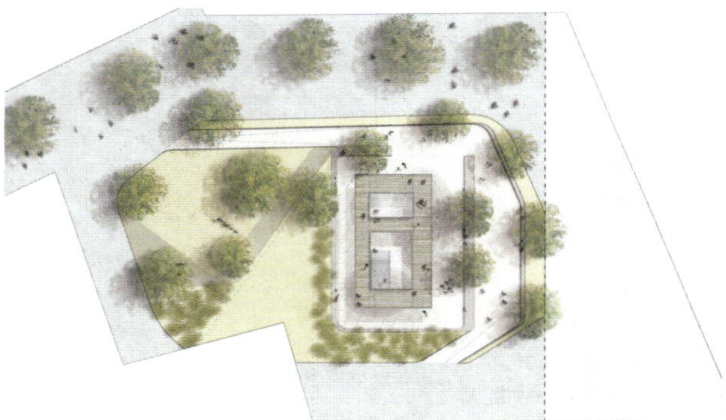

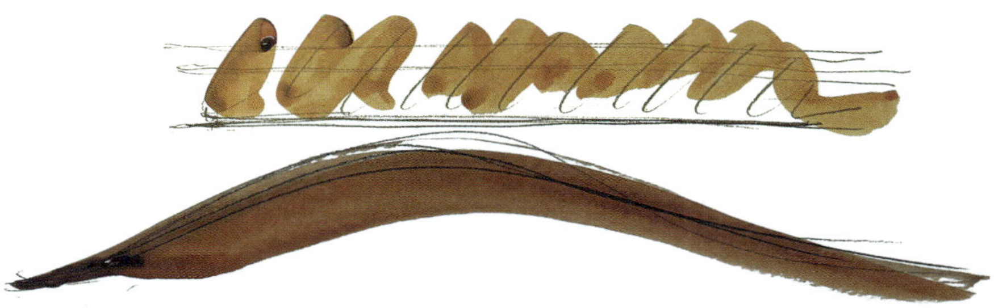

Santa Maria Mazzarello Plaza Landscape Game

The architects wanted to rehabilitate a lost urban space and create a social and recreational meeting point with an intervention that consists of a large children's play area around which a series of paths, planters, and green areas are articulated. For children, formally recognizable playground equipment, such as a castle or a small house, often lead to repetitive games that do not stimulate their creativity. Therefore, the project, on the other hand, is committed to establish a play landscape in which children can give free reign to their ideas and imagination by varying the shape, surface, and material of the intervention, while also adding traditional play equipment, such as slides or nets, to the levels. In the new park environment, under the shade of the trees, the benches and seats serve as an anteroom to the centerpiece of the garden.

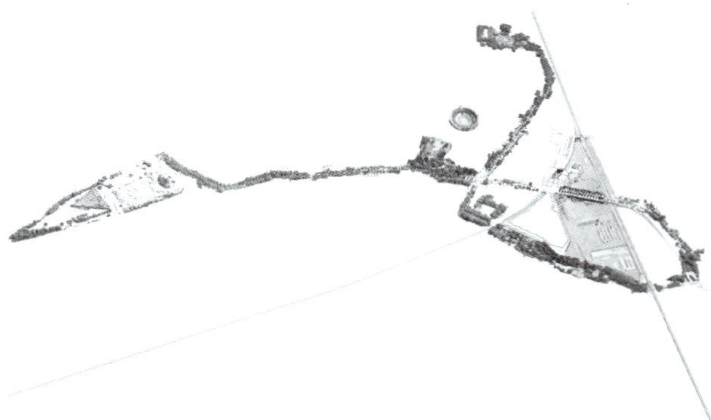

Luma Parc des Ateliers

Luma Arles is an experimental cultural center situated in the former industrial railyard of Parc des Ateliers. Bureau Bas Smets transformed this wasteland into a welcoming park through the recolonization of 80,000 trees and plants of the abandoned site. The design considers what nature would have done over time. Dunes of fertile soil would have been blown there over time by the prevailing winds. Pioneer plants would invade the shallow soil depths, while an ecological succession would occur with more available soil. The vegetation, combined with a sustainable water system consisting of a large pond, produces significant evapotranspiration, resulting in a micro-climate that transforms the former concrete semi-desert into a Mediterranean climate. The project does not re-establish a lost past but expedites the future transformation.

Eco-district Victor Hugo in Bagneux

The town of Bagneux, with almost 39,000 inhabitants, is located two kilometers southwest of Paris. Due to the extension of line 4 of the Paris Metro and its connection with line 15 of the Grand Paris Express, the Victor Hugo Eco-district, which revolves around this future major public transport hub, anticipates and accompanies the coming changes, the transformation of the city, and the new ambitions of the Bagneux area. The central objective of the ongoing project is to make the station a central point of the district and to support job creation and economic development. The aim is also to create a controlled density around the station to enhance social and programmatic diversity and connect the neighborhoods more closely. Emphasis is also placed on highlighting green infrastructure and integrating environmental concerns.

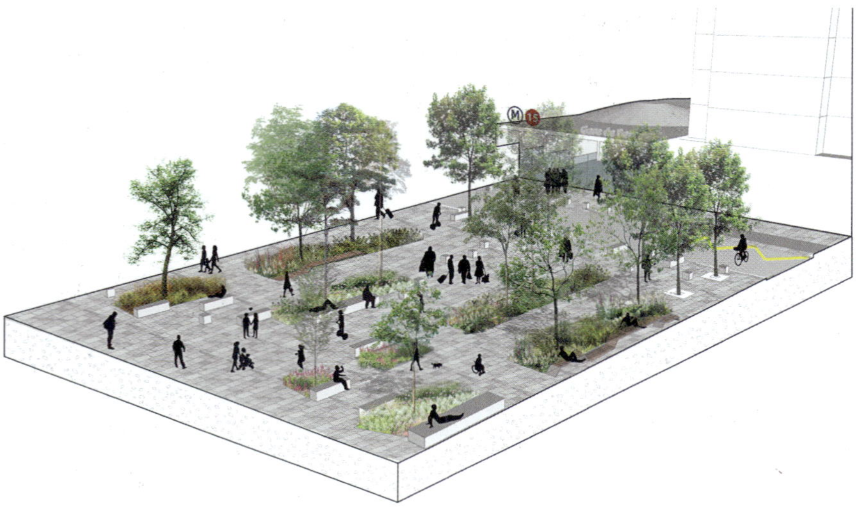

Nature's Forgiveness (International Garden Festival)

To celebrate the 25th anniversary of Des Grands Jardins et Paysages, DGJ was chosen to create a garden for the International Garden Festival of Chaumont-sur-Loire, which attracts 500,000 visitors each year. At the heart of the garden, an architectural composition initially leads the eye to a cultural garden, with man-made varieties of plants. Inside the garden, a wild environment reveals itself, with natural ancestors of our cultivated plants, such as meadow sages and wild roses. The broken bridge symbolizes that in the process of recreating nature, mankind lost its connection with nature, even though it is dependent on the symbiosis between everything on earth. The small garden wants to let nature take over again, as a reminder that nature is ultimately forgiving.

All Strelitzias Go to Heaven

The project All Strelitzias Go to Heaven was designed for the 28th edition of the International Garden Festival in Chaumont-sur-Loire, France, with the theme Jardins de Paradis. For the designers of this garden, which won the Prize for Creation, paradise is a landscape with blue skies, fluffy clouds, and extraordinary plants. They therefore created a garden in which the boundaries between sky and garden are blurred and the green of the living leaves blends with artificial polygons of clouds. This paradise is planted with plantains (Musa paradisiaca) and strelizia, also known as the bird of paradise flower. They embody all the characteristics expected of plants: spectacular flowering, bright colors, delicious fruit (in case of the plantains), elegant leaves, or an attractive fragrance.

Les Jardins du Montperthuis – The Montperthuis Gardens

The garden, designed by Philippe Dubreuil, was initially an empty expanse with a few cows, so the first action was to work on the topography: therefore a terrace to the 16th-century barn, a green house, an enclosure wall to protect the formal manor house courtyard, and several basins were created. An avenue of elms leads to the east-facing manor house and a fountain with four water jets at the front gardens center. On both sides of the main axe, rooms of hornbeam host collections of hydrangeas, peonies, and ferns. A wrought-iron gate separates these floral rooms from the vegetable garden, at the center of which a rill provides water for the vegetables. An orchard completes the needs of the household. A rose garden with a collection of more than 400 ancient perfumed roses connects sprawls between the well and the greenhouse.

Les Jardins d'Etretat

Perched atop Normandy's windswept Alabaster coast, the Jardins d'Etretat, which cover 15,000 square meters, emerge as a breathtaking fusion of nature and artistic innovation. Unfolding in seven thematic parts, the site invites visitors on a sensory journey conceived by landscape architect Alexandre Grivko. Through his vision, the gardens unite the classical methods of André Le Nôtre with surrealist topiary shapes. The contrasting topography is creatively addressed through architectural solutions, while the shapes of plants recall the natural phenomena of Normandy: cliff arches, English Channel waves and sea shells. In this project, Alexandre Grivko carefully prioritized the historical integrity of the site and the beauty of the local ecosystem.

Esplanade TASE

The project realizes a public green promenade next to a former listed silk factory called TASE factory and new residential buildings located in Lyon. The construction was based on the principle of circular economy. For example, the excavated material was reused for the work in the public space. Various measures were implemented as part of the project. Gravel was produced as well as roadway layers and pedestrian path substructures. Other excavated material was used for drainage mats, earth and stone mixtures, and new seating. A soil management plan was used to diagnose the soil quality and quantity. The project created a positive atmosphere, practical seating framed by greenery, combining efficiency with a particular focus on sustainability.

Redevelopment of the Place Jean Jaurès

This redevelopment reclaimed and renovated the once neglected and unused Place Jean Jaurès in Marseille to provide a public space for many pedestrians. The project's objective was to create a large, vibrant, versatile, and popular Mediterranean city square. Furthermore, the uniqueness of the La Plaine district was preserved by taking a contextual approach to the human and cultural environment of the neighborhood. A multitude of trees frame the square and create a pleasant atmosphere where people of all ages feel comfortable and are able to come together.

Place Nationale Montauban

The Place Nationale of Montauban, whose arched walkway has been listed as a Historic Monument since 1910, has been completely restored. An action plan has been drawn up to enhance the historic square: surface treatment, lighting enhancement, creation of a water mirror with misting devices, organization of terraces for cafe owners and restaurateurs, all within the context of a sustainable development strategy. The 700-square-meter water mirror with over 300 misters, can disappear to create a mineral surface throughout the square, right up to the covered areas on all four sides. Particular attention is paid to the street furniture designed for the square. Benches and pots are adorned with snakes, motifs borrowed from the collections of the Musée Ingres Bourdelle, just a stone's throw away.

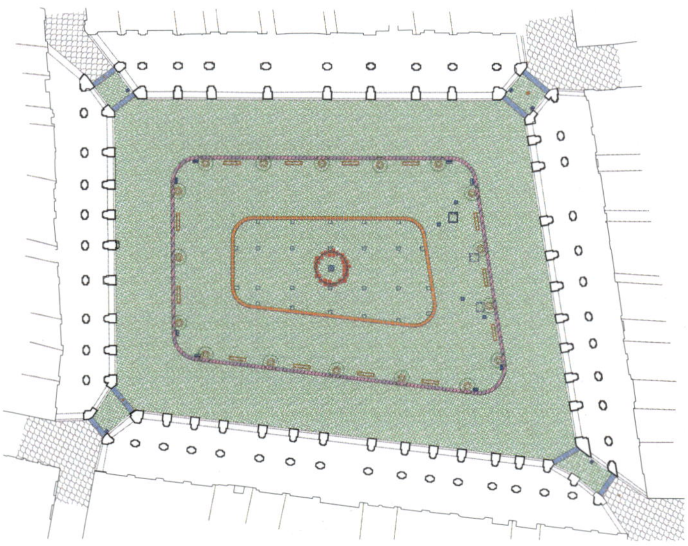

The Arches

The Arceaux district in Montpellier is an ideal place for children, except for the confusing public space that insufficiently separates motorized traffic from soft mobility. Base landscape agency responds to this problem by taking up the theme of city at children's height, linking the city center to the Arceaux district. As a precursor, Collectif Parenthèse offers a temporary entertainment space that allows a gradual transformation of public space by giving children a place of honor: a class outside the walls with a grandstand and two large tables on the forecourt of the Chaptal School, swings, a motor skills trail for the little ones, a dance mirror, a fog field, a mini-petanque or dairy court, and large picnic tables. The modules can be assembled, disassembled, stored, and reused or used in outdoor events, festivals and school playgrounds.

Promenade Nardal

Located in the 14th arrondissement of Paris, the Promenade Nardal is an experimental and ambitious project that combines density, nature, and culture. The promenade is located on the site of a former hospital and above a tunnel of obsolete railway infrastructure, which is home to the largest urban colony of the pipistrelle bat, a protected species of bats, in Europe. The spaces host different programmatic functions based on the discovery of biodiversity and conviviality: the Kid's Lounge offers games for children from three to twelve years old, the Discovery Lounge is a place to organize events between neighbors the Games Lounge offers a journey around the world and Greater Paris, the Sports Lounge is a kind of open-air sports hall, a shared garden, and an amphitheater for events.

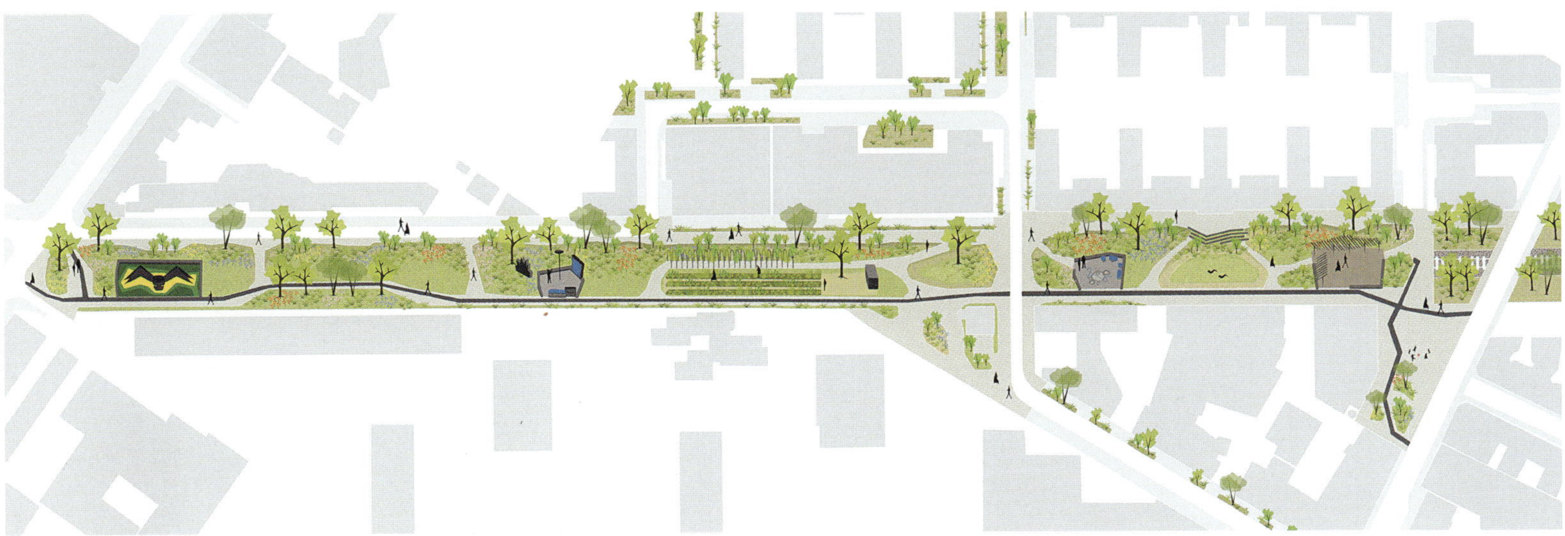

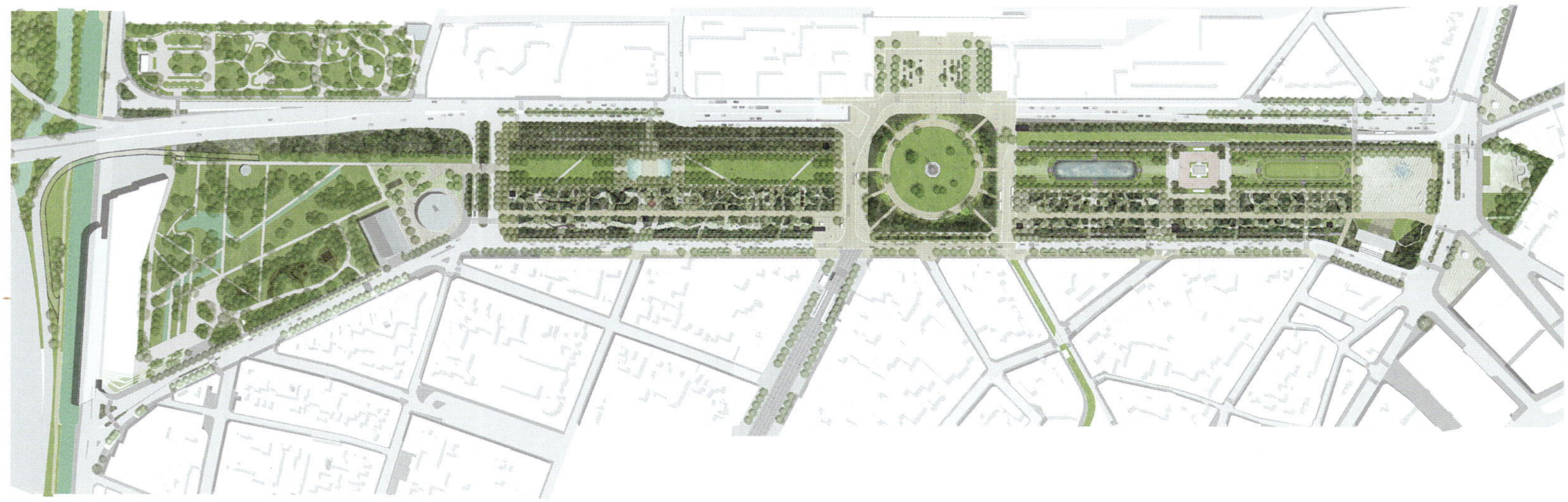

The Promenades of Reims

Established in the 18th century on the site of the moats of the ancient ramparts, the promenades of Reims connect the historic city center to the suburbs. In spite of additions along an axis established and disrupting infrastructure they remained unified by a dense tree heritage composed of chestnut trees, maple trees, and plane trees. The city wished to transform it into a "Central Park à la rémoise" of 15,000 square meters. The project removes invasive parking from the promenades and restructures by thinning on one side, while complementing the tree heritage on the other to accommodate new playful and friendly practices. As an urban forest it offers atmospheres reminiscent of forest woodlands. Furniture promotes a multitude of uses, and the promenades are rehabilitated as spaces for socialization and collective well-being.

Topomotif

Topomotif was created through a series of design-build workshops with children and parents in the city of Romainville, outside of Paris. Landscape architects Sensomoto translated these ideas into a unique public play space, encouraging free-play and a positive sense of risk. The artful playground brings kid-size scale and a new identity to the urban environment. Sensomoto teamed up with artisan C-Cube and safety compliance Prélud to craft the custom structures. Natural materials are maximized: the structures are wooden, trees provide shade, planting areas bring biodiversity and fall zones are covered in bark mulch. Brilliant yellow metal and rope accents add distinction. Nine unique play modules, specialized for the needs of kids with different ages and abilities, simultaneously accommodate a variety of users and usages.

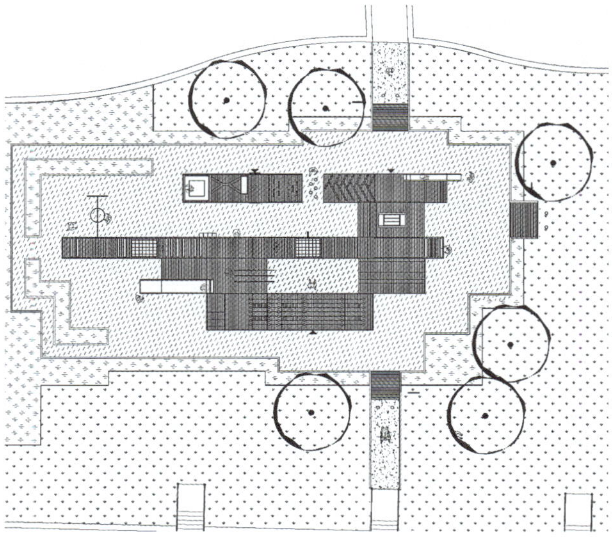

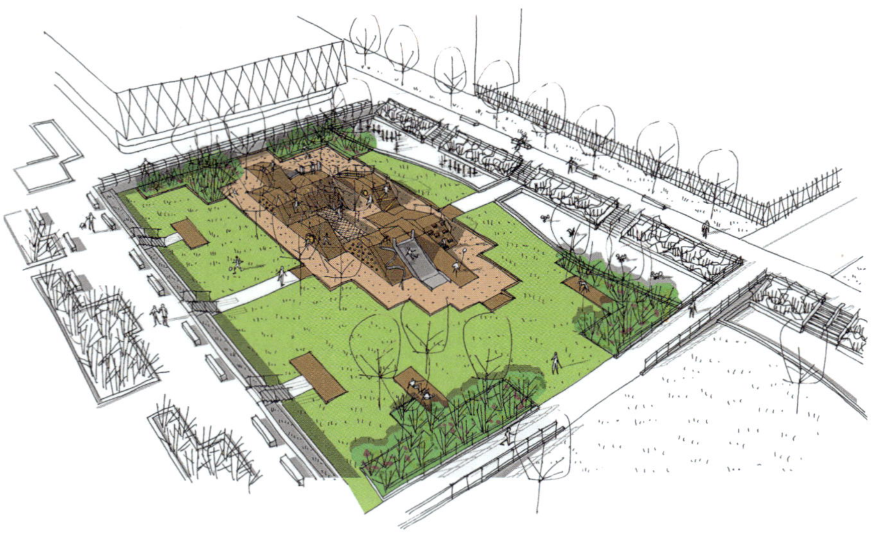

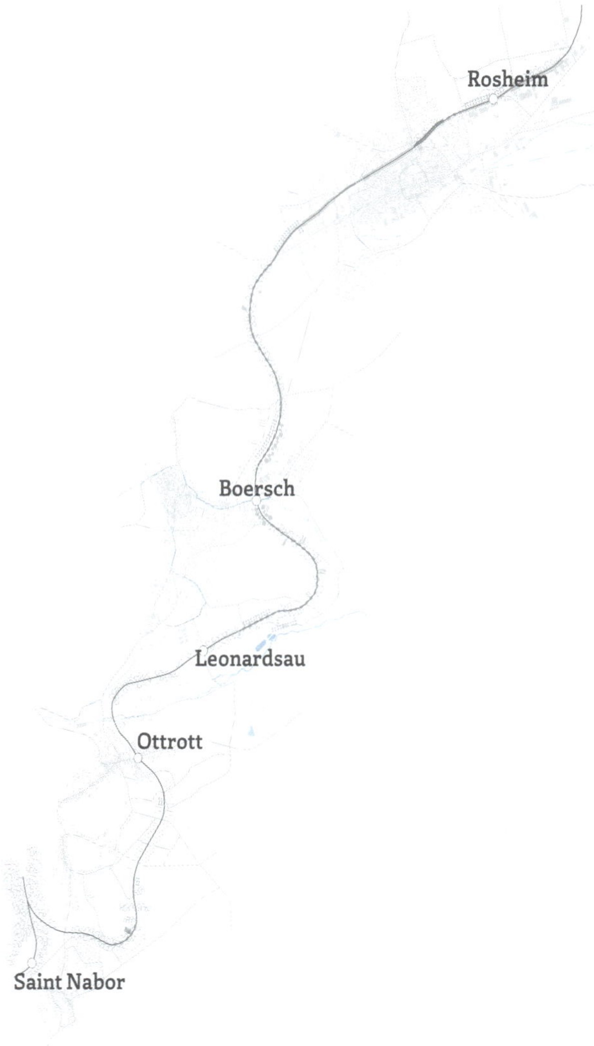

Chemin des Carrières – The Path of the Quarries

The Chemin des Carrières is a lace undulating in the landscape that invites people to travel. The ominous vestiges of the railway line still mark the reading of the site. A story is told along the 11-kilometer-long path, which is divided into five chapters at the stops. They cover different sections of the landscape and places, in which there are repeated encounters with water. The desire to create a route for the quarries had to adapt to the undulating landscapes of the sub-Vosges hills and tells the history of the landscape and the men. The journey of discovery through forgotten or everyday landscapes is intended for both locals and tourists. Like the old railroad line, which had a dual function, industrial and passenger transport, the route has a dual vocation where the functional must rub shoulders with the imagination of travel.

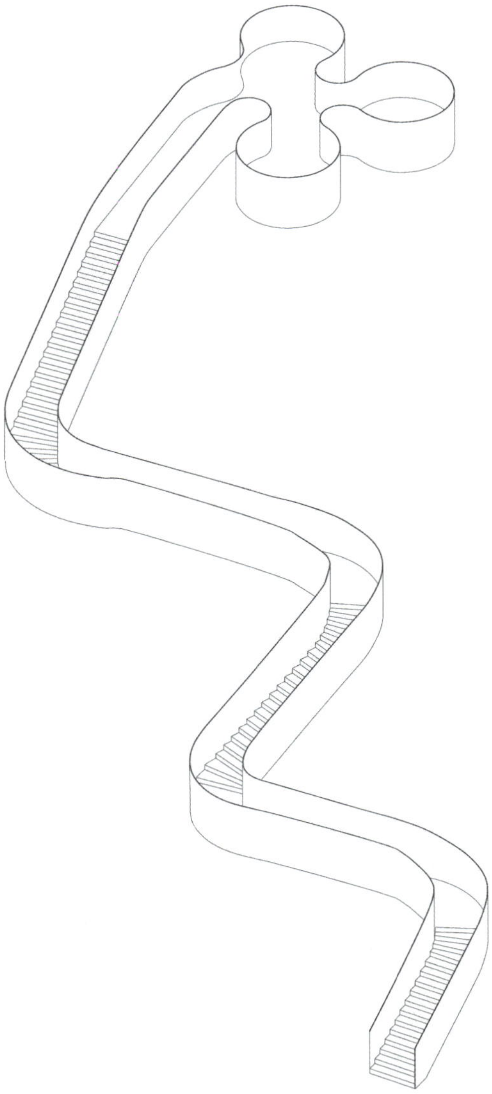

Belvédère of Port de Rozé

The Port de Rozé, located at the interface between the Brière marshland and the inhabited islands, is a stopping point that invites visitors to enjoy a bird's eye view of the island landscape. The project includes reducing the size of the streets, converting the existing parking lot into a permeable, planted parking lot, renovating the old café and creating a pedestrian zone, and a new square. A 24-meters-high Belvédère made of a metal structure with wooden cladding and floor provides a viewpoint to admire the landscape. A digital periscope, installed on the ground floor, allows a barrier-free access to the augmented reality panorama. The wooden canopy offers a lively and animated space for locals and tourists alike. The renovated riverbanks open onto the Brivet river, recreating the town's historic connection to the water and navigation.

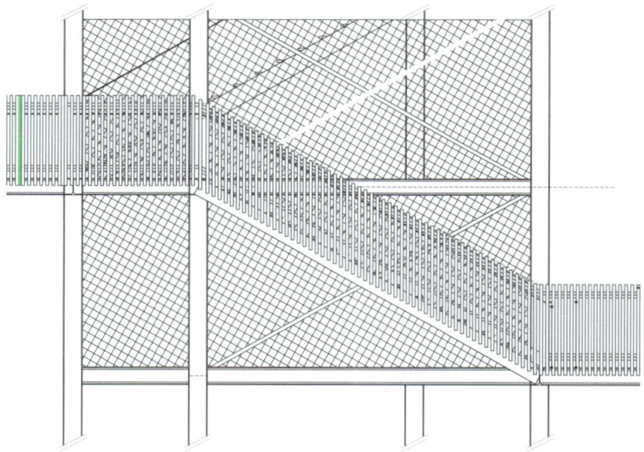

Parc de la Corniche des Forts

The park is part of a network of green spaces stretching over three kilometers of hillside in the towns of Lilas, Romainville, Pantin, and Noisy-le-Sec, East of Paris. This site, a former gypsum quarry, has been transformed into a nature park consisting of a 200,000-square-meter inaccessible sanctuary for the existing vegetation. It is surrounded by a 5,000-square-meter promenade open to the public, which bypasses the central forest and overlooks it. This allows the ecosystem of the renaturalized quarry to thrive while highlighting it and providing leisure and sports facilities for the community. A pedestrian bridge – 124.70 meters long, three meters wide and 6.50 to 8.50 meters high – brings pedestrians closer to the forest. Viewing huts are nestled above the bridge's two piers, inviting contemplation and discovery.

Resurgence of the Petite Saussaie Spring

The project brings back an historic wellspring, that was previously buried and piped but now flows openly, following its course over 150 meters in the public space of Vitry-sur-Seine. The descent of the water in the Seine valley is the subject of a sculpture-fountain and an aqueduct. The aqueduct raises the water within sight, hearing and touch. The water remains horizontal as often and as long as possible and, conducted by the aqueduct, leaves the ground of the hill several times. winds its way through landscapes of dozens of willows, wet gardens, aquatic plants, elongated basins or pebble shores. The multiplicity of places crossed are unified by the continuous gravity flow of water from the Petite Saussaie Spring. By the hydraulic system, the places are brought into tune with the geographical scale.

Multi-generational Square in Rodange

The low-traffic square between the school center and different sports facilities creates an extension of the playing and communication space for children and the town's citizens. Vibrant urban culture finds a fertile place to flourish. As a meeting place for the generations, the square offers space for funfairs, trade fairs, and markets. Shady, tree-lined tiers serve as rest, observation, and meeting areas for citizens. In contrast, an urban wall stage serves as an activity and play zone for young people. Loose rows of trees give the square a shady, green frame. The mixture of tree species creates contrasts for an experience for all senses. The paving of the square is replaced by infiltration-friendly surfaces towards the edges.

The Crack

The Crack is located in a cooperative garden city designed by the pioneer of modern garden cities and landscape architect Louis van der Swaelmenin in Woluwe-Saint-Lambert. The project reclaims a former tennis court that has not been used by the community for a long time. By exposing the ground, the project renews its life cycle and creates new social life. It features social sculptures created from found materials. The Crack creates an intriguing third space as an invitation to the inhabitants to also become active. The valley connection wished by Van der Swaelmen is reinvented by opening the fence, the ground and the hedges, establishing a playspace. Today, the Crack ends against a gate closed by the public owner of the site. By emphasizing this in their design, Taktyk and Alive Architecture criticize this absurd decision.

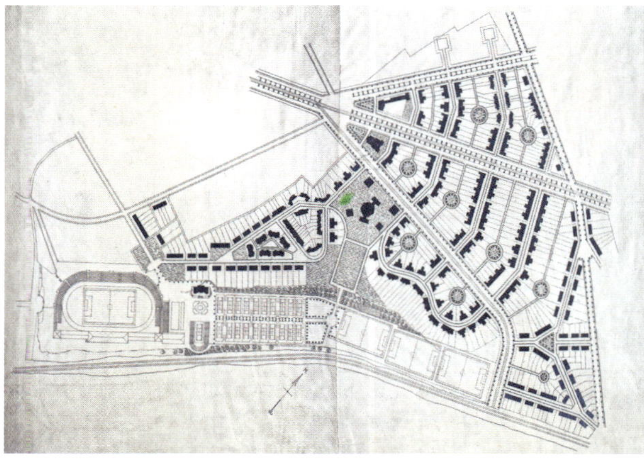

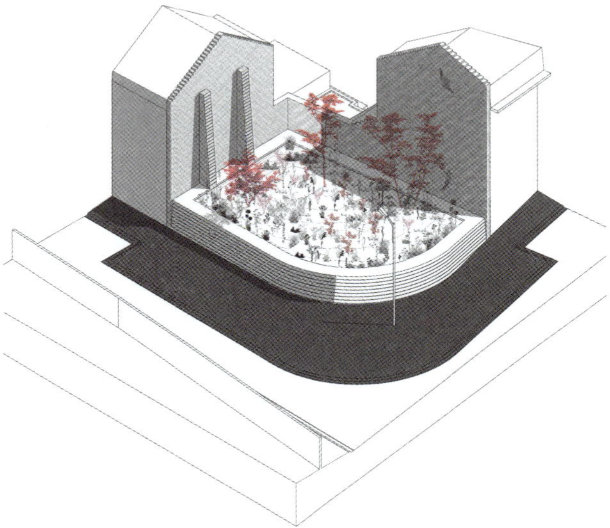

Memorial Garden between Heaven and Earth

The Memorial Garden between Heaven and Earth succeeded in transforming this highly charged site in the city of Charleroi, where kidnapped kids were held captive, into a meeting place, paying tribute to the victims. Today, the space is brought back to life with a place of remembrance, a marvellous garden that gives way to color, softness, and light. The garden offers local residents a green space, with colorful vegetation that evolves throughout the year according to the seasons. White, pink, and violet flowers populate an evergreen carpet. The ensemble brings together local and more exotic species, including three trees: a Davidia involucrata, a Cornus kousa, and an Amelanchier lamarckii. The flowerbed contains a variety of plants, including grasses, perennials, and bulbs.

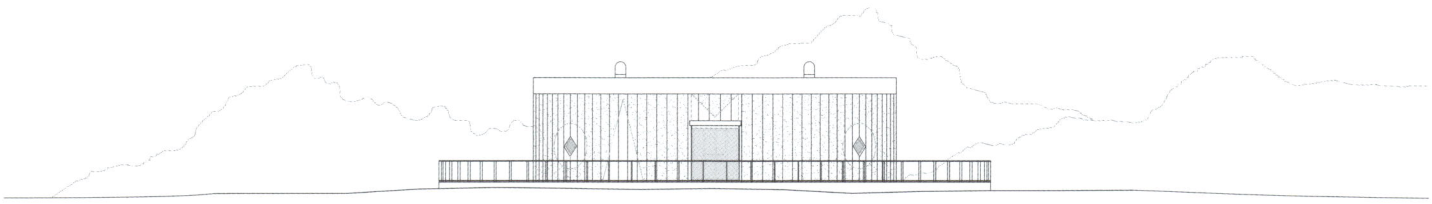

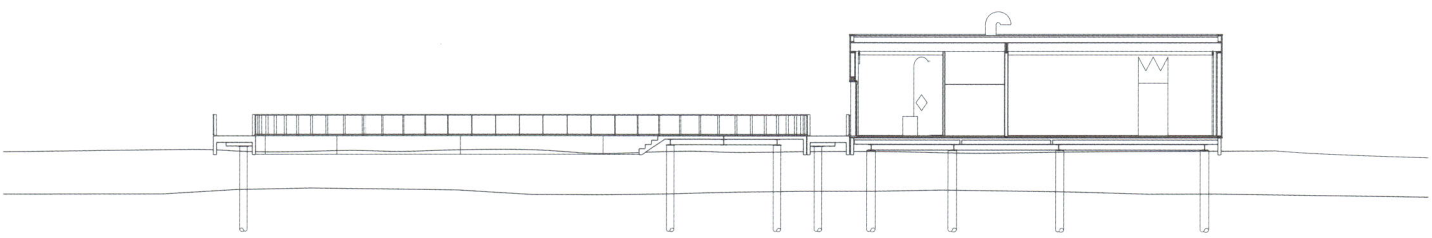

Art Pavilion M.

The Art Pavilion M. – floating on the water of the Weerwater – has been realized in the course of a competition, enhanced by the Dutch city of Almere and a museum for land art. The project includes a sequence of three tangential rings: the port, the stage, and the observatory. The port includes a water square where art is exhibited and birds and plants can be observed. The stage, a second circle, is the terrace of the museum within the ring. Along the ring, it becomes an outdoor terrace. On special occasions, it is used as a stage on the water. The observatory is a lightweight structure that encloses the exhibition rooms and serves as a greenhouse for the surrounding vegetation. Enclosed by a curtain, cut-out shapes let the light through in different ways. The roof is covered with shells that filter the rainwater and provide a buffet for birds.

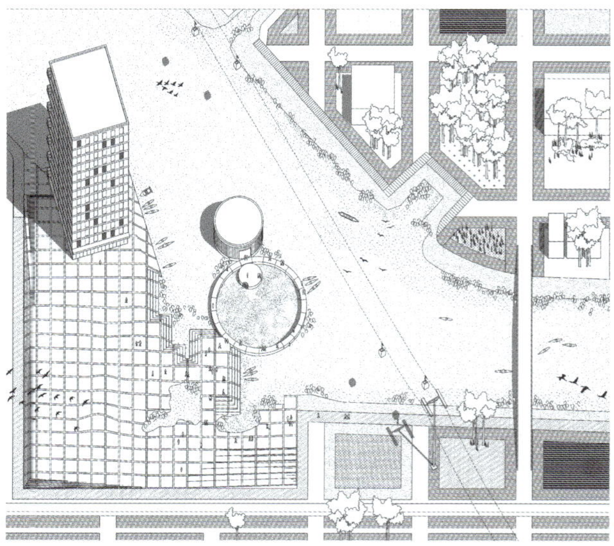

Twickel Estate

Twickel, whose history goes back to the fourteenth century, is the largest private estate in the Netherlands. Among others, Strootman Landschapsarchitecten has drawn up a renovation plan for the Overpark, based on respect for the historical park design, adjustment to the new circumstances, and a focus on future use. The Overpark was opened up, with the renewal of groups of trees and the addition of others here and there. To achieve greater cohesion between the different parts of the Overpark, an elongated open space has been created in the woodland to connect the open area in front of the castle with the woodland by means of a beautifully curved space. Strootman designed a new pond and seven new pedestrian bridges: six in Overpark and one over the ring road to reinforce the connection between Twickel and Delden.

Schoonschip

This renowned circular neighborhood was initiated and developed by a group of enthusiasts with a shared dream: to build a sustainable, close-knit community on the water. For more than a decade, the members of Schoonschip have worked to create their dream homes. The smart pier connects all the houses to each other and to the necessary technical infrastructure, and is designed to facilitate casual encounters between members of the community. A plot passport ensures a sense of unity despite the architectural diversity of the houses. Schoonschip residents live in highly eco-efficient, remotely manufactured buildings and share everything from electric cars and cargo bikes to the clean energy generated on the houseboat roofs.

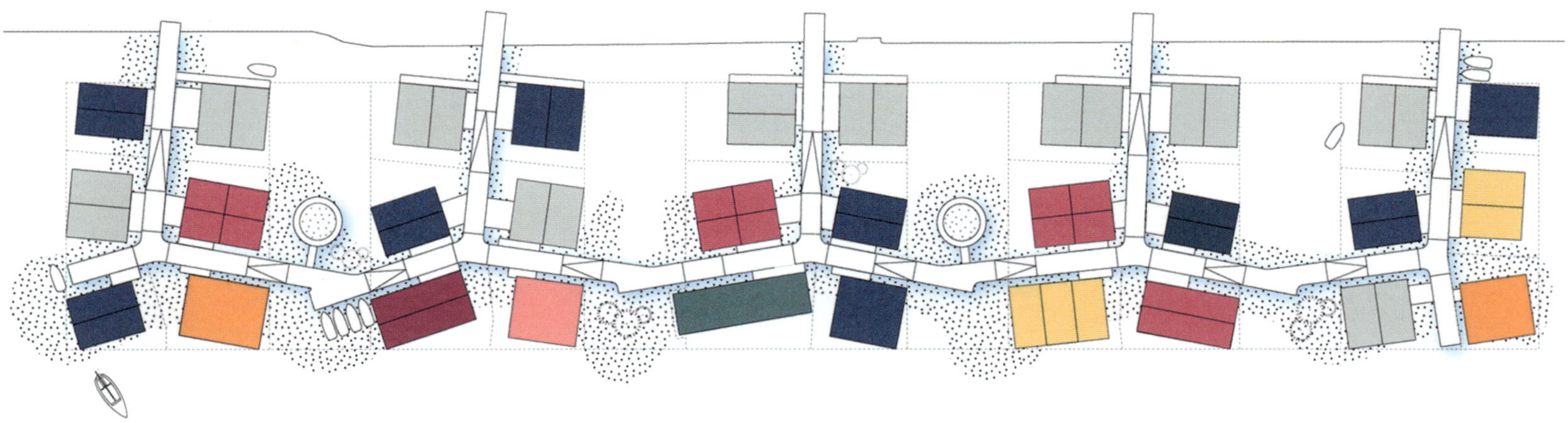

Aan het IJ Amsterdam

A new neighborhood of 1,850 dwellings has been developed on the northern bank of the IJ river in Amsterdam, containing multi-level luxury apartments embedded in a calm, green urban landscape. The Dutch landscape architecture studio Buro Sant en Co has designed the public space of the streetscape, parks, squares, and twelve remarkable collective courtyards. Each has its own signature style expressed by unique objects or furniture, planting, and staying quality. The green public environments are primarily designed for slow traffic, where pedestrians and cyclists are mixed with local traffic in minimal paved street profiles to ensure maximum value of green streetscapes. Next to attractive squares and parks, the neighborhood also has a unique experience of its surrounding context by water related shores and parks.

Peel Plaza

Created with artist and designer Gabriel Lester of studio PolyLester, the public space Peel Plaza by Capital C Amsterdam forms a work of art that invites physical interaction. A wooden decking whose parts appear to have been peeled off incorporate functionality, sculptural forms, six benches and tables, as well as existing trees and greenery. The peeled-off parts, where neighbors or passersby can sit or lie down, form lively and spatial gestures on Weesperplein in the eastern part of the canal belt of Amsterdam. There is also the opportunity for visitors to have lunch or drink to stay: Capital C's grand café called Capital Kitchen will be open during the terrace period to provide hospitality.

Ede Market Square

The historic green character of Ede was the inspiration for the renovation of the Market Square. New, modern Overtuinen – detached gardens across the street – with hedges and trees were reintroduced by Hosper. The designers had to deal with the constraints of an underground parking garage, the surrounding road and railway, and the space required for the market and events. The Overtuinen are slightly elevated and surrounded by hedges and flowering plants, giving them their own environment for playgrounds and terraces. The square has a pleasant size and topography, with attractive small water fountains in the center, which are often used by children. In this way, the square is a pleasant and popular place for meetings and activities in a green environment.

Fifty-Fifty Villa

In collaboration with Studioninedots, The Dutch Wave has succeeded in creating a unique villa in which the open and closed spaces alternate. Hence the name of the house: Fifty-Fifty. One half is closed and the other half is open. The interior and exterior spaces of Villa Fifty-Fifty are equally important and merge into one another. Generous windows give the feeling of being part of the surrounding nature. At the same time, various plants have been integrated into the architecture of the building. A clear example of this is the tree that grows from the ground to the roof. The Dutch Wave designs gardens according to the biophilic principle and looks for ways to reconnect with our natural habitat in every design or concept. As can be clearly seen in the Fifty-Fifty Villa, nature is always in charge.

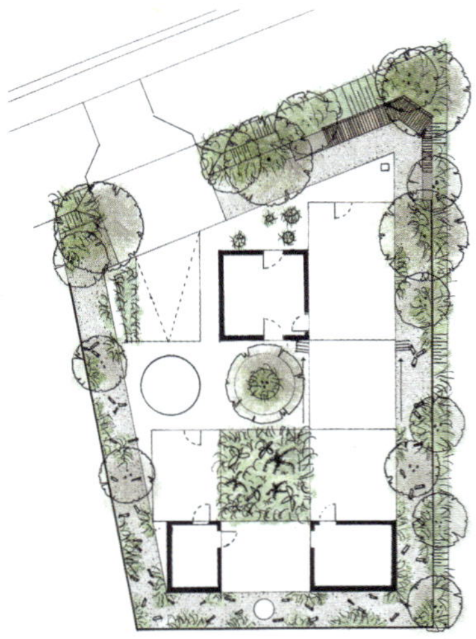

Gardens De Hooge Riet

Former sanatorium De Hooge Riet – now national heritage – was recently transformed into a residential area. The close relationship between the buildings and the historic park and gardens was restored. In the forest belt around the complex, H+N+S Landscape Architects focused on conserving the atmosphere of the forest. The newly added buildings in this area – designed by diederendirrix architects - are in close harmony with both the existing trees and the historic building ensemble. In the park area planting beds, pergolas with climbing plants, and some original sight lines were brought back. In collaboration with Jacqueline van der Kloet, a planting plan for the old and newly added courtyards was developed that preserves valuable trees and historic elements while accommodating the new function.

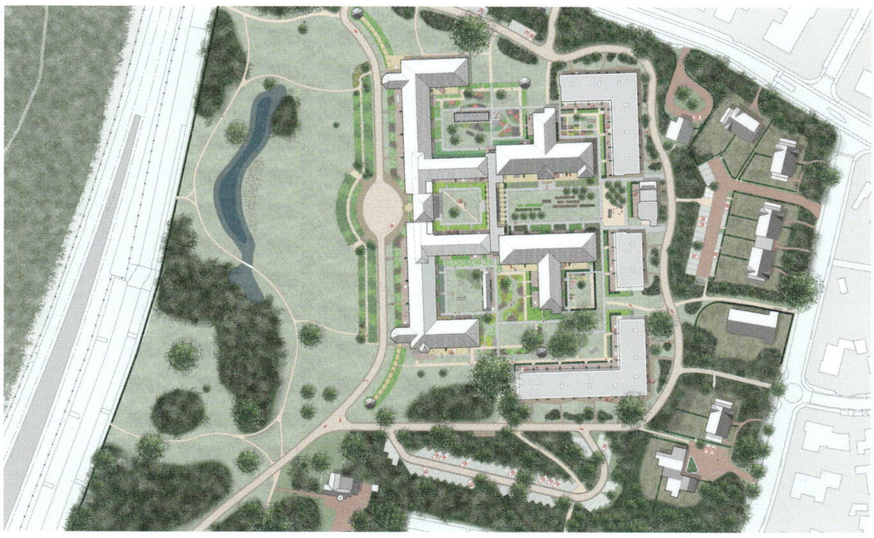

Loenen Memorial Cemetery

The Loenen cemetery is located in a nature reserve on the Veluwe, consisting of forest and heath. The cemetery was established in 1949 on the initiative of the Dutch War Graves Foundation (OGS). The cemetery is a triangular area with a wooded character, with room for approximately 5,000 graves. Karres en Brands developed a design for the new cemetery, which will be implemented in several phases over a long period of time. The extension features an autonomous walkway with a series of elliptical open spaces. The path, a carved passage with heather vegetation, leads from the main entrance through the forest to the current cemetery. The open spaces are separate burial areas, each with its own character. The position of these spaces is related to the topography, the surroundings, and the existing cemetery.

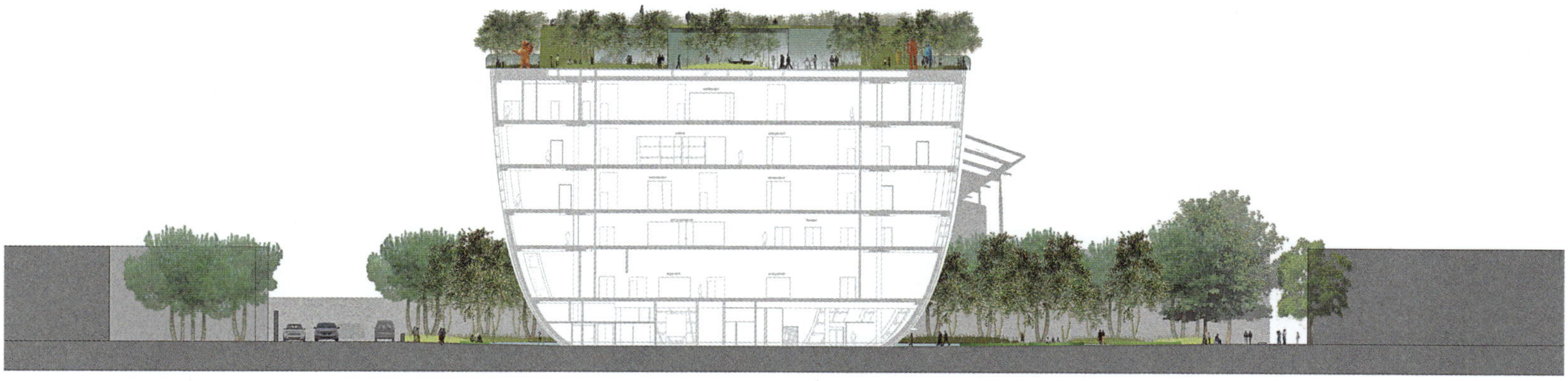

Depot Boijmans van Beuningen

In the middle of the Museum Park, MTD landscape architects, in close collaboration with MVRDV, have designed a special public meeting place around and on top of the new Boijmans van Beuningen Depot in Rotterdam. With the Depot, Boijmans van Beuningen is the first museum in the world to offer access to its entire collection. The public square and the rooftop forest contribute to this openness. The design of the square and roof top gives the surroundings the form of a green forest world. The plan with a fragmented structure extends from the forecourt to the roof landscape 32 meters above. Together they form a robust and cohesive landscape. The vegetation of Pinus nigra, Betula pendula, and Stinzen plantings gives the visitor the impression of being in an urban forest.

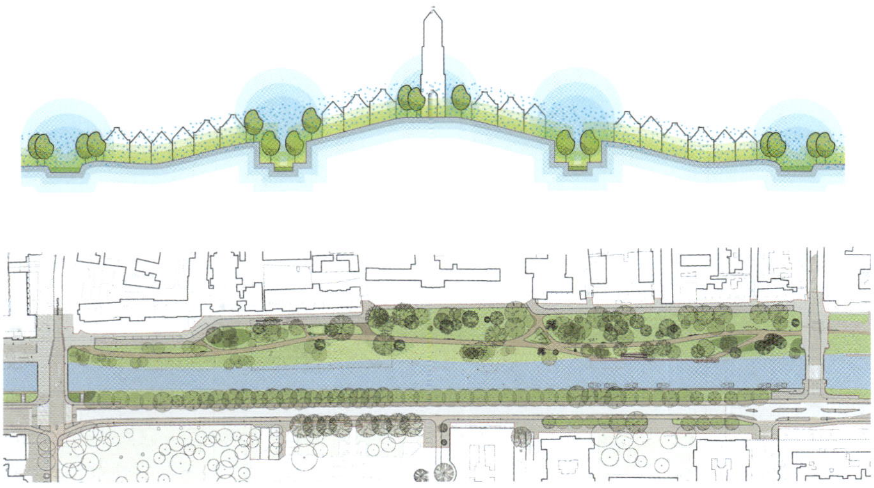

Catharijnesingel in Utrecht

The project Catharijnesingel is connecting the historic city and the station, improving the livability and walkability of the public space, with significantly less car traffic. Around 40,000 cubic meters of water have returned to the now six-kilometer-long canal. The extensive walking route along the canal invites recreational and sporting use. To focus on nature-inclusivity, the park contains grasslands, conversation pieces, numerous seating areas, and a variety of greenery such as evergreens and richly flowering species. Materials like old Rhine bricks are used as baked clinkers and create a visual connection with the historical inner city of Utrecht. A wooden deck has been placed next to the existing jetty, which can be used as a seating element, a stand, or a stage. The park has become a comfortable place to stroll, stay, and to enjoy.

Het Platform – Microcity Microlandscape

Het Platform is a mixed-use building designed as a MicroCity by VenhoevenCS, featuring a complex series of functional roofs. LANDLAB provided these outdoor spaces for recreation, while also stimulating biodiversity, buffering water, and improving the surrounding air quality, making it one of the first nature-inclusive residential buildings in the Netherlands. The planting concept is inspired by the vegetation in the different landscapes of the province: plants inspired by allotments on the first floor, the undergrowth of the estate forest on the third floor, orchard meadows on the sixth floor, and a mountain birch forest on the eighth floor. The plantings are specially selected to provide optimal conditions for attracting pollinating insects. The goal is that, over time, birds will feel safe enough to stay and even reproduce here.

Bridgefoot Street Park

Bridgefoot Street Park is a unique spatial composition using construction and demolition debris to create much-needed open space for people to interact with the natural world in Dublin's city center. The designers engaged in a creative process of community engagement with local gardeners, early school leaver trainees and members of the prison after-care service to develop the layout. Concrete, stone, and brick material waste was used to create landscape finishes including paving, retaining elements and topography. Materials were also crushed and mixed with sub- and topsoil to create different substrates on which vegetation would naturally colonize over time. Many tonnes of what would have been waste was creatively redirected away from landfill, the embodied energy now permanently stored within the new park.

Regents Place

Regent's Place is a mixed-use campus, comprised primarily of office space. The public realm design focused on improving the quality of the experience through inclusive design principles, and major greening for health, wellbeing, and biodiverse sustainability. A retrofit scheme, complex aspects were considered, including load balancing, constrained depths, and basement ventilation. The team consulted local communities to develop a shared vision, which incorporated significant areas of new planting, paving, and street furniture, to create a habitable, attractive public realm where people from all walks of life can sit, play, and engage as part of a site wide promotion of healthy and active lifestyles. The project has transformed Regents Place from an uninviting corporate aesthetic into a welcoming green inclusive public amenity.

Urban Playground

The Urban Playground is neither a sculpture nor a traditional playground. It is inspired by wooden children's toys and is abstract yet brings an element of unexpected play into the city. Larger pieces have cavern-like spaces, where smaller modules have been 'carved' out. These create the opportunity for seating, open-ended play, or exploration. There is no right or wrong way to play here, nor is there any explicit narrative to follow. The Urban Playground is deliberately eye-catching and is laid out as a 'trail of breadcrumbs' that leads visitors and people passing by on a journey through a linear pedestrian street. Materials including expanded natural cork were chosen for their multisensory appeal, ease of fabrication, and environmental sustainability.

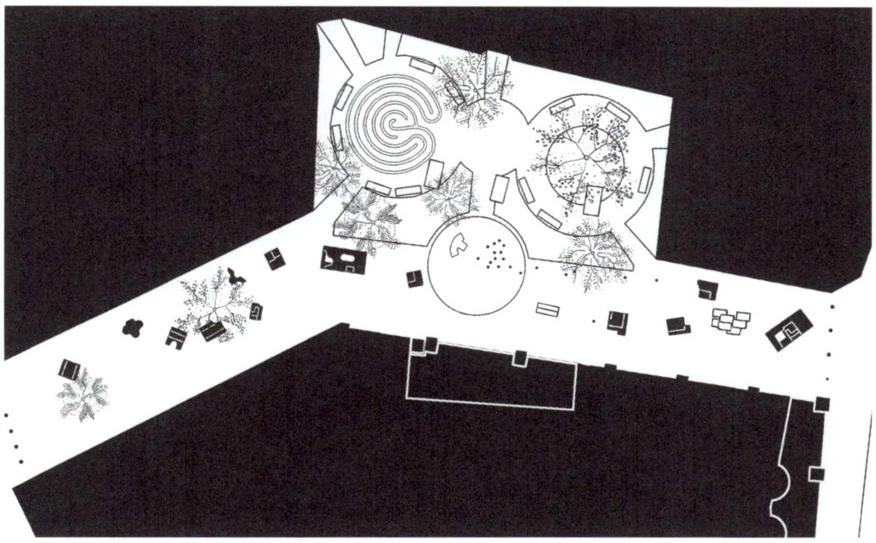

Moor Lane Community Garden

In an effort to improve the amount of green space in Londons Moor Lane, a new temporary installation called the Moor Lane Community Garden has been created. The project was developed through a series of collaborative workshops together with the neighborhood, and explores innovative approaches to greening our streets, while responding to the history and design of the region, drawing inspiration from landmarks like London Wall. In addition to increasing greening, the project serves as a trial of elements of a future landscaping program for Moor Lane, which will strive to provide further biodiversity and tree planting. After the permanent program is in place, the planters will be moved to different public areas in the neighborhood, leaving a lasting impact on the community.

Superbloom at The Tower of London

Designed to mark Queen Elizabeth II's Jubilee, Superbloom turned an historic but underused moat into a spectacular floral display. It aimed to promote nature's value and provide a rich city center haven for visitors and insects. Superbloom featured a planting scheme of over 20 million seeds selected to deliver visitors a different experience every visit. This new planting required importing and laying of an entirely new soil profile above the existing ground level. Several installations enhanced the experience, including a large multi lane slide, a series of immersive gardens along the east moat including a collection of copper insects, lighting, soundscape, and a collection of bespoke glass artworks. Superbloom rises to the challenges of climate change, hoping to inspire similar projects to be as impactful in the future.

Alfred Place Gardens

As the first new park in London's West End for more than 25 years, Alfred Place Gardens replaces a dull street dominated by vehicles. Reimagined by LDA Design as a calm oasis and richly planted linear park, the street now prioritizes pedestrians and cyclists. Local people wanted somewhere to sit and a place that was soft and green. So, the redesign features long sculptural, elegantly curved seating, ideal for chatting, as well as playful elements to attract families. Biodiverse woodland planting nestles under a mature tree canopy, and there are lounging lawns and an event space bordered by grasses and colorful perennials. Alfred Place Gardens is a key part of Camden Council's award-winning West End Project, which was designed to improve air quality and health, biodiversity and access to green space.

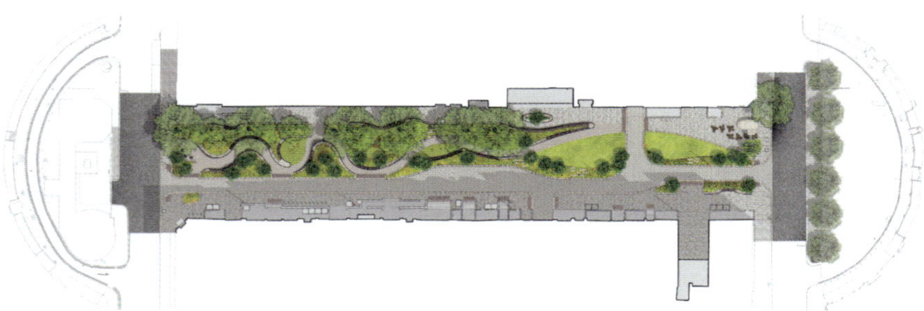

PRICE&
MYERS

Fidelity Heart Zone

The site, at the heart of Fidelity campus, looks to renew a tired and unloved space to create a thriving green heart, which encourages connections with nature and social interaction between staff. The landscape design realized a more fluid and contemporary setting for staff and visitors. Elements include a car drop-off point, new pedestrian link from the car park, a clear hierarchy of routes, and a sequence of arrival for visitors. The Central Heart contains a planted rain garden to capture surface water run-off from the adjacent surfaces with generous seating steps, specimen tree planting, and relocated artwork to provide focal points. Detailed planting reflects different landscape typologies, such as hillsides, productive gardens, and woodland planting, to provide a diverse and biophilic setting.

Mayfield Park

The 65,000-square-meter project includes around 1,500 apartments, one million square meters of office space, a 350-bedroom hotel, and retail and leisure facilities. The park consists of a series of character spaces connected by a meandering river. The design strikes a balance between actively usable green spaces for people and natural habitats for wildlife. The river Medlock was previously culverted, but uncovering the river has played a key role in the reintroduction of wildlife to the area. A naturalistic planting concept has been implemented throughout the park with different plant arrangements, planting beds, and lawns that can adapt to changing climatic conditions. Mayfield Park features open lawns, riverside paths, three new bridges over the River Medlock, a playground, and a variety of seating areas.

Crusader Works

Capital & Centric commissioned the redevelopment of Crusader Works, a garden for residents, to create flexible social spaces within a lush, green setting. The ideas of sanctuary, rhythm, and formality were taken forward and playfully overlaid with a materials palette that responds to the industrial heritage of the building. The planting comprises a grove of multi-stemmed Parrotia persica complemented by groundcover planting that draws inspiration from the self-seeded plants that were found growing in cracks and crevices in the pre-restored Mill. The boundaries between hard and soft landscape have been blurred with grass creeping through the paving and trailing plants cascading from window boxes. Communal seating, fire pits, and barbecues allow the community to socialize in their space.

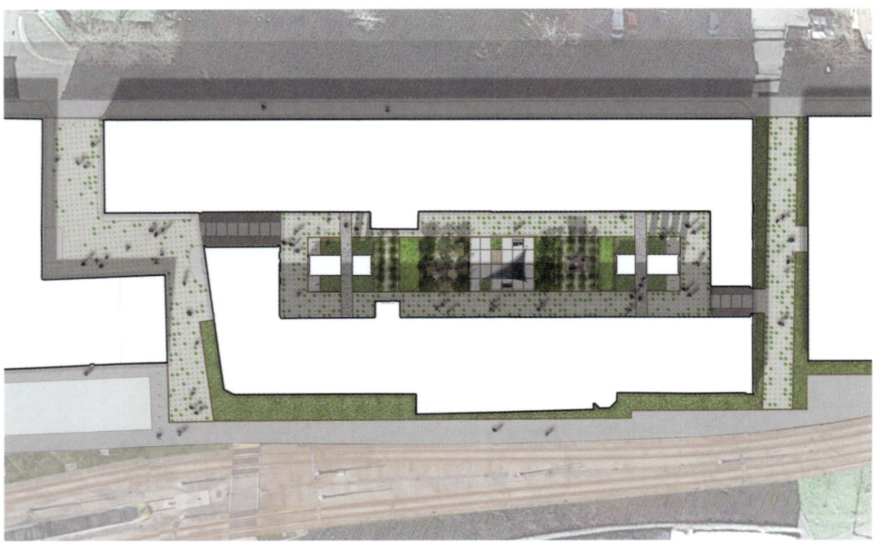

The Viper

The Viper is a tree canopy walkway at the Newt in Somerset, UK – a country estate with splendid gardens and major facilities. The walkway curves through a working woodland on a steep slope and links a large meadow with a Museum of Gardening. The steel structure, with its spine beam, was prefabricated in components in South Africa and shipped to site in open containers. All was erected by crane and bolted together onto steel tripods on pile footings, carefully positioned to avoid major trees. The forest was laser scanned in three dimensions, so that the curving structure would fit the faraway site. The walkway is 130 meters long and 1.5 meters wide, with a wider section for visitors to linger and admire the views throughout the changing seasons.

The Bowline

The Bowline is a new active travel route and elevated linear heritage park that revitalizes the disused Bowling Viaduct and swing bridge. The travel route eliminates a dangerous road crossing by creating an off-road section of National Cycle Route Seven connecting Glasgow to Loch Lomond and the Trossachs National Park. The design of the 700-meter-long route is comprised of several distinct character areas that correspond to the surrounding landscape context and the three viaduct sections. The linear park contains accessible viewing platforms, blocks of woodland planting; a hierarchy of paths for walking and cycling and boardwalks, and stepping stones for children to explore. The materials compliment the industrial character of the structure, including pre-cast concrete, cast iron, and powder coated steel.

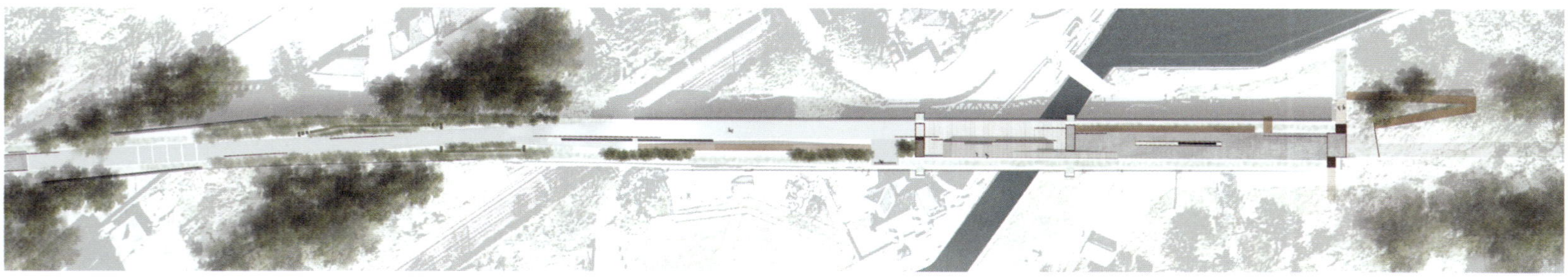

1763
1769
1790
1768
1775

Aalborg Waterfront

The master plan for Aalborg Waterfront connects the medieval city center with the adjacent fjord. The qualities of the approximately one-kilometer-long waterfront are emphasized by a tree-lined and unusually detailed boulevard for cyclists and pedestrians. A harbor promenade with steps and recessed terraces allows people to get close to the water. Various types of urban gardens facilitate activities such as markets, ball games, and sunbathing. The central activity area is designed to accommodate a variety of games and sports, from beach volleyball in the summer to an ice-skating rink in the winter. The area is surrounded by dramatically angled nets and light poles. The materials chosen are asphalt, rubber, weathering steel, concrete and wood, with subtle references to the sea through wavy paving patterns.

Carl Nielsens Allé

Previously, Carl Nielsens Allé was a busy parking lot. Now, the space has been transformed into a lush urban area, which the local residents can enjoy every day. During heavy rain, the green oasis turns into a beautiful water basin, effectively managing stormwater in the area. With the Copenhagen Municipality as the commissioning party, Carl Nielsens Allé has been transformed into a vibrant urban space, which combines nature, city life, and climate adaptation in a charming neighborhood with a simple architectural intervention. It is an oasis in the middle of the city where traffic, environmental, and urban challenges have been transformed into a strong, sensory, and lush urban space, intensely green and consisting of a high and a low-lying area surrounded by a multifunctional edge furniture.

Ikea Copenhagen

Urban life, biodiversity and the best views of Copenhagen await visitors on the roof of this Ikea store. 20 meters above ground 1:1 Landskab has created a green sky park– open to the public – with wild urban nature and room to relax. The project spans three floors with forecourts and urban spaces on the two lower levels, giving access to the store and bicycle parking and a rooftop garden on top. The garden is inspired by the Danish cultural landscape and country-side – with a strong focus on biodiversity. It contains three different types of nature: the beach meadow, the pasture, and the hedgerows. The planting consists of robust local species that grow naturally in windy, nutrient-poor biotopes. They provide shelter on this windy roof, thus creating a green environment for humans and insects alike.

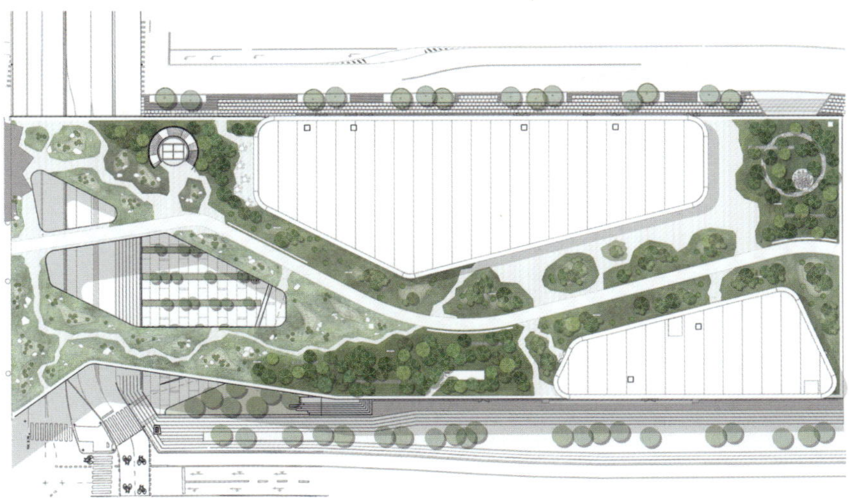

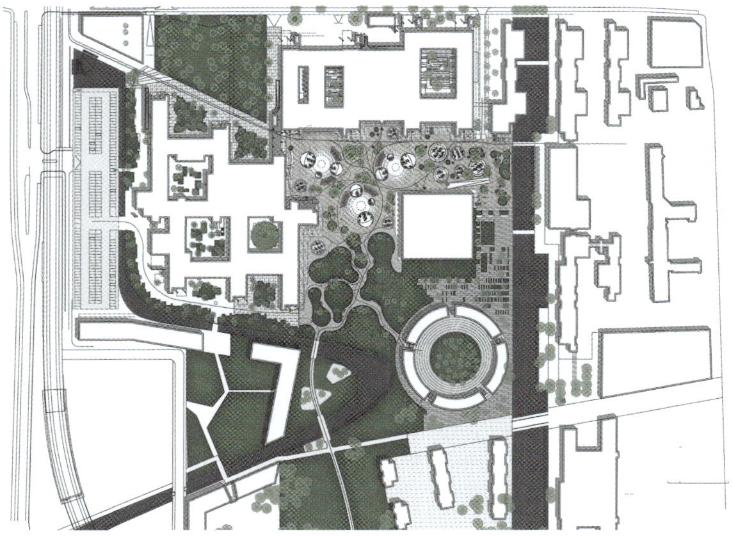

Karen Blixens Plads

This 21,415-square-meter urban plaza at the University of Copenhagen is one of the biggest public squares in Copenhagen. The plaza is located between the recently built university buildings of the South Campus and the nature reserve Amager Commons. The urban space is connected to Amager Commons through a gradual transition, which makes Karen Blixens Plads a hybrid of park and public square, organized as a superimposed surface of human-made hills and valleys with room for 2,000 parked bicycles inside the hills. The square works as a campus landscape with an important functional role as well as a recreational resource. The necessary infrastructure is turned into a three-dimensional student hang-out.

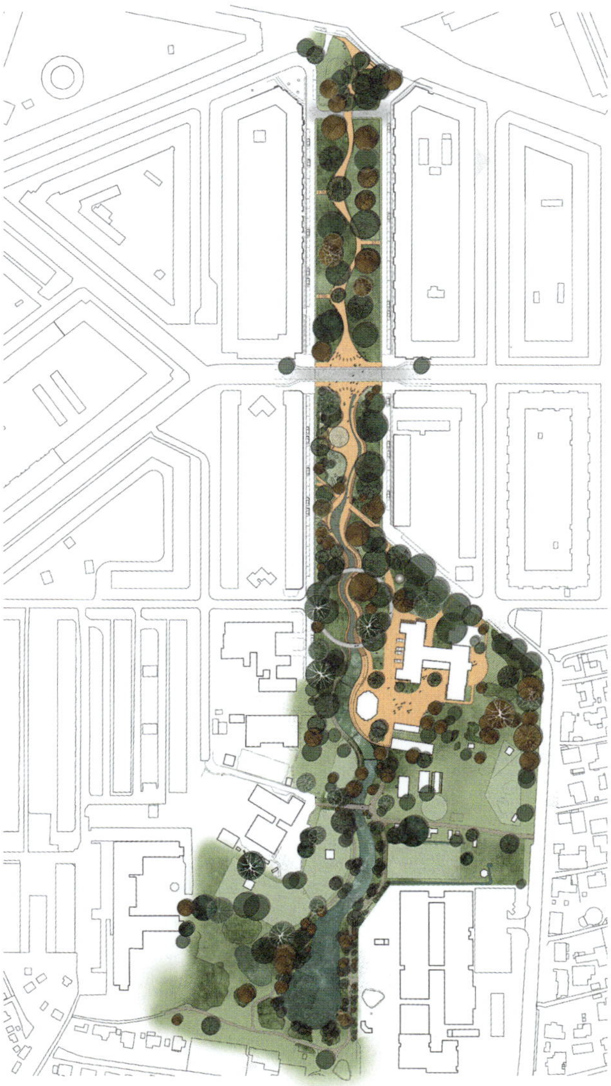

Karens Minde Axis

The cloudburst and urban space project Karens Minde Axis in the Copenhagen area of Sydhavn combines retention of up to 15,000 cubic meters of rainwater with the development of parks, urban spaces, and nature. A tiled path wavers in and out between the many large old trees. In the event of a cloudburst, water is transported through a 600-meter-long course from Ellebjergvej in the north, past Karens Minde Culture House, and is finally collected in a recreational rainwater basin. The project aims to improve the physical framework of the positive development of Sydhavn, where a unique grassroots spirit is paired with culture, nature, and wildlife. A large part of the area's cultural and everyday life as well as social activities will unfold in the Karens Minde Axis, a green connection between Vestre Kirkegård and Tippen.

→ Vega Landskab → Interventions in Nature → 2022 → Lemvig → Denmark → Photos: Anne Dorthe Vestergaard, Arne Overby

Lemvig Sø

Lemvig Sø is a glacial lake situated in the heart of the town of Lemvig. By opening up the lakeside, the water has become accessible for everyone and integrated with the rest of the place as a cohesive space by establishing new gravel paths. The central part of the project, as a universal design, is a new seating staircase and ramp that gives access to the water's surface. The fluctuation of the water levels offers unique experiences throughout the year. A climbing net has been incorporated so that, at normal water levels, one can hover above the surface and get a close sensorial experience with the water. Native plants along the lakeside restore the area back to its natural habitat, creating a welcoming environment for wildlife species.

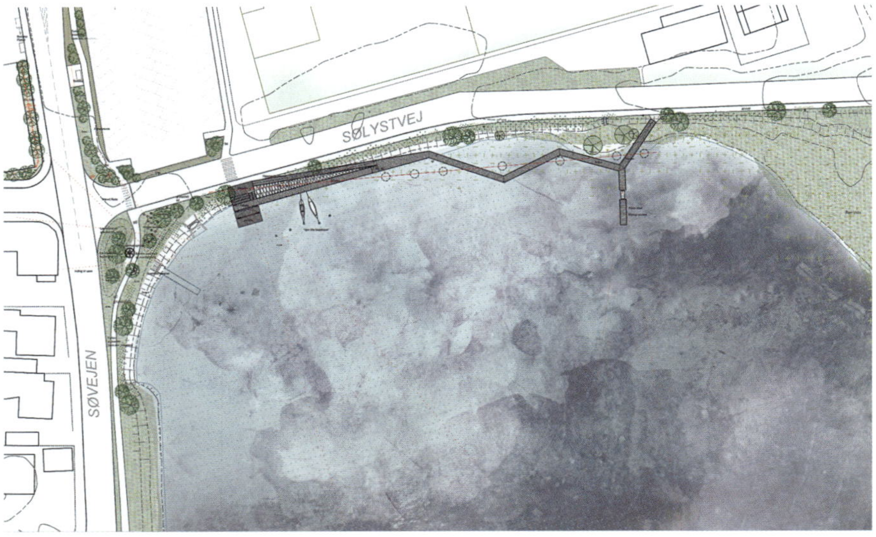

Skovbrynet Basecamp

Skovbrynet Basecamp, by landscape architects Kragh & Berglund, and architect Lars Gitz, is part of a lively green area in Lyngby, surrounded by Sorgenfri cemetery to the north and Lyngby lake to the west. It is the location of 640 new apartments for students, researchers, and seniors. An organically winding network of paths on the roof allows visitors to experience the view over the lake. The planting is dense and park-like with scattered trees, robust plants, and cut and tall grasses. The roof contains a lush landscape with private terraces for the residents of the building. The shape of the building helps create interesting spaces along the exterior façades, as well as a unique inner courtyard with little niches and recreational areas.

Ballerup Boulevard

Bicycle lanes, paths, and flower beds have transformed Ballerup Boulevard from a congested traffic route into a revitalizing green zone. Space for cyclists and pedestrians has been created by reducing the boulevard from four to two lanes. By relocating bike lanes and pedestrian paths away from the roadway, the safety for users improved. The old lanes have now become a recreational park with open LAR water treatment, trees, and vast flower beds. The greenery creates small spaces with benches along the pavements and bicycle lanes. The boulevard is the backbone of a larger urban development and will serve as an important green corridor in the long term.

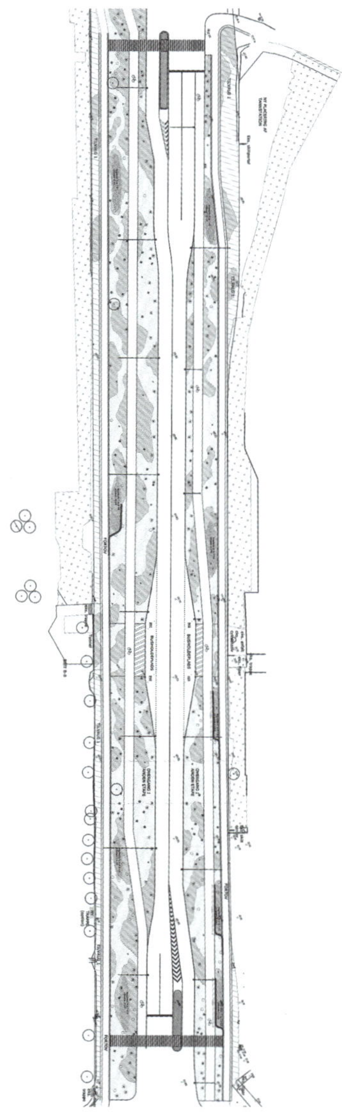

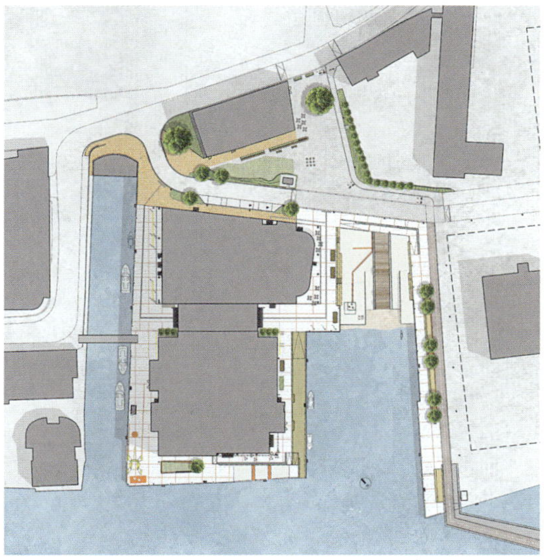

Slippen Marineholmen

At Slippen in Møhlenpris, a new meeting place has been created by establishing a new urban beach where the slipway of the old shipyard was historically located. The urban beach has become an attractive gathering point in the neighborhood that helps to connect the historic built environment in Møhlenpris with the business park at Marineholmen. The area is used for temporary markets and pop-up events. Outdoor seating from multiple cafes, along with activities like skating, sunbathing, and swimming, has transformed this area into an attractive year-round meeting spot for the residents. It serves as a diverse hub for community gatherings and events. The urban space encompasses the beach, its edges, the undulating waves, steps, and wooden decks, all offering attractive options for leisure and recreation in the city.

vestre

The Plus

The factory and visitor center for furniture manufacturer Vestrev designed by BIG – Bjarke Ingels Group – is located in a 1.2 square kilometer park in Magnor, Norway. It doubles as a public park for hiking and camping, and is in keeping with the region's mission to create a green manufacturing hub outside Oslo. The Plus is conceived as a radial arrangement of four main production halls – a warehouse, paint factory, wood factory, and assembly hall – that connect in the center and create the Plus shape at their intersection. The central hub wraps around a public, circular courtyard where the latest outdoor furniture collections are displayed and changed with the seasons. The radical transparency invites visitors and hikers to enjoy the whole process of creation, while providing Vestre's team with the thrill of working in the middle of the forest.

Voldsløkka Secondary School Sustainable Courtyard

For this sustainable school, the schoolyard was adapted for climate change particularity. Biodiversity and stormwater management complement the usual functions of a schoolyard. Encircling the islands are lush, vegetated channels designed to collect rainwater from the surroundings. These channels have the capacity to become completely filled with water. In the aftermath of heavy rainfall, the water is either absorbed by the vegetation or gradually filtrated through the soil. Crossing over the channels are universally accessible bridges, crafted from grids, hovering above the vegetation. By using stormwater as a leading design parameter, the landscape architects created a unique schoolyard, with a lush, green character, and great importance of sustainability aspects.

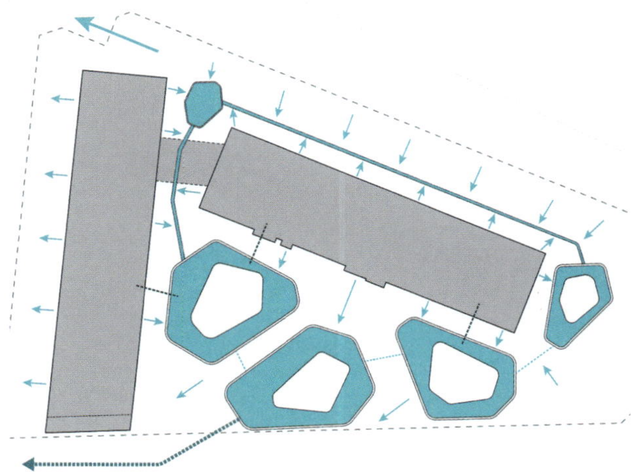

Armadillo

The Armadillo project is part of an experimental landscape design and land art project called Strandhugg, initiated and curated by the Swedish-Norwegian design studio Byggstudio. Rintala Eggertsson Architects, together with two other studios, were invited to submit ideas for the revitalization of a large coastal stretch on the southwestern coast of Sweden. With the installation of Armadillo, a small shelter was created, partly to guide people away from the most vulnerable areas with heath biotopes, but also to provide shelter from the wind. The installation is a simple wooden structure that can stand on its own and has the capacity to accommodate up to six people. The Armadillo is anchored in the sand with earth-spears, allowing the structure to stand firmly and to be moved if required.

Broparken in Linköping

Project Broparken is an engineered stormwater system, disguised as an attractive urban social space, transforming a characterless stretch of land into a much-loved public park, featuring the stunning red bridge, developed with designer Johan Kauppi. The communal urban space includes meadows, fruit trees, and a stream. This popular urban park is an important part of the ecosystem, providing habitat for otters, herons, and pollinating insects. Broparken is the first project in the new urban development Vallastaden. The ideal environmental conditions are taken into account at an early stage so that flora and fauna can thrive better. Robust plants and building materials have been chosen to withstand the fluctuating water levels. The park invites community interaction by providing a variety of accessible green spaces and landscaped areas.

Torparängens Park

In an area of overgrown, historic farmland, the project explores the topic of re-breaking fertile soil while elaborating the concept of organization versus natural processes. The central pathway leads the visitor through several strata of human time scales: from culturally ancient meadows with prehistoric tombs, via pre-modern farmlands to contemporary landscape statements, becoming a gesture to observe both the historic site, the representations of landscape, and the labour invested in organizing plants and grounds. The park emerges in its geometrical order, borrowed from the structure of agriculture anchored to fragments of a preexisting stone wall. The articulation of order gradually appears towards the north where the grid-planting stands out, in contrast to the naturalistic south part where trees are planted right on the path.

→ Photos: Jussi Murole, Maanlumo, Marko Kallio, Skyfox

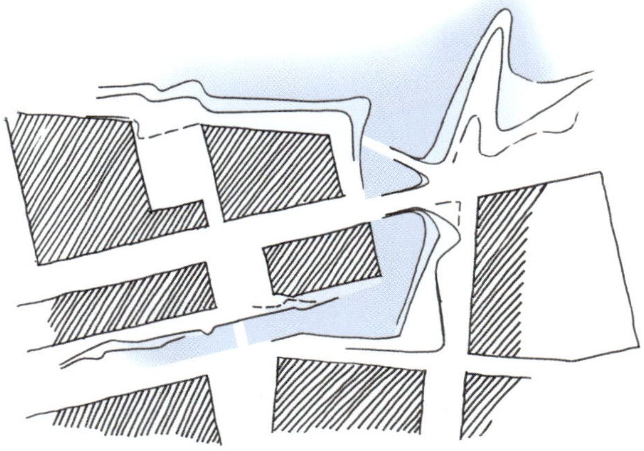

Ranta-Tampella Public Outdoor Spaces

This new residential neighborhood is located north of the city center linking it with Lake Näsijärvi and nature. A water canal connects the lake with the historical Tammerkoski channel and rapids. The shapes of the bridges, shorelines, and canal banks imitate the movement of the waves. The design of the public spaces seamlessly connects canal banks, squares, streets, parks, shores, docks, and recreational routes and is based on the intertwining of the new urban and the natural environment. The vegetation is designed with natural and dynamic aspects, and recycled materials in, for example, pavements and shoreline construction. The city plan stems from B&M architects; the outdoor space design was done with Pontek, Aihio Architects, Suunnittelukide, Ramboll Finland and Lightning Design Collective.

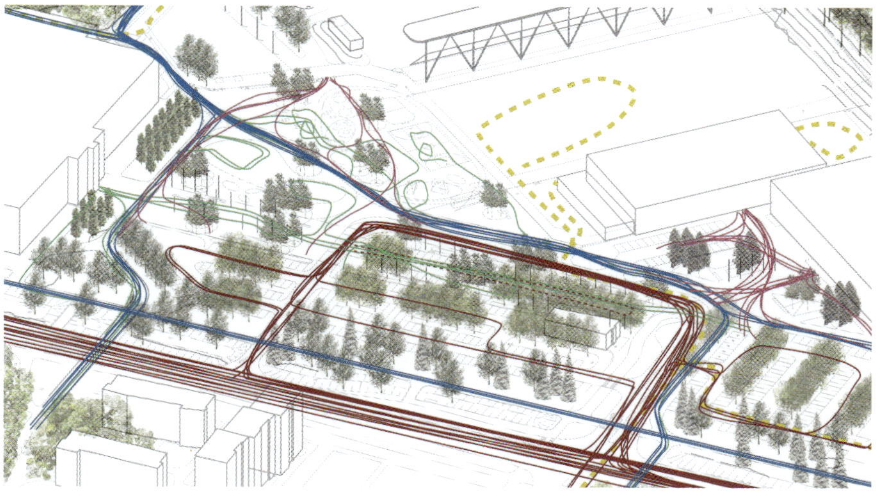

Myyrmäki Sports Park

Myyrmäki Sports Park is a multi-purpose public sports center in the city of Vantaa. The sports park is located in a prime location in the heart of the larger sports complex. Previously, the park had traditional sports facilities, and the redevelopment aims to provide a public exercise area, with new functions open to all, making the joy of movement visible. The overall look is cheerful, fresh, and green. The focus is on providing a safe and accessible exercise area for all ages. Therefore, island-like areas with different sports facilities are placed between the access paths. Through reorganization, the park has found the potential of an existing site. Play and exercise extend beyond the designated sports area: the temporary parking structures, all the benches and drift barriers are designed for skating.

Hermannsplatz Bad Muskau

In the municipality of Bad Muskau, Hermannsplatz was created on the site of the former border checkpoint. At the particularly sensitive interface between the town, the banks of the Neiße, and the Unesco World Heritage Site Muskauer Park / Park Mużakowski, the site has been transformed from a formerly car-dominated transit area into a lively square. The newly designed Hermannsplatz is divided into a garden square section and a traffic-oriented area that takes into account the historical layout of the square. The traffic-calmed square area with contrasting strips of natural stone paving and concrete planters gives the site its individual character. The generously spaced planters are planted with tall oak trees. Towards the Neiße, a clear spatial edge sets the scene for the bridgehead on the Neißedamm.

Garden with Lawn Sculpture

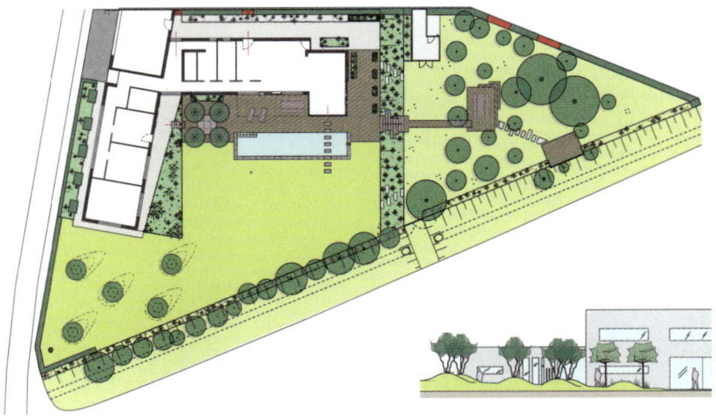

The main design elements of this leisure garden are a house terrace with an ornamental pool, a seating area integrated into an existing orchard, and – as an eye-catcher – a sweeping lawn sculpture. The house terrace with a smooth-planed wooden deck includes smaller planting islands and a dominant shady seating area formed by four canopy-shaped sweetgum trees. A wooden walkway leads from the deck through the rustic orchard to a spacious seating area amidst the old gnarled apple trees. An eye-catcher of a special kind is a gently swaying lawn sculpture with five umbrella-shaped rock pears at the crests. The lawn sculpture fills the slightly recessed niche at the western end of the property and transforms it into a secret highlight of the garden, which effects the perception from every corner of the extensive garden.

Pool as focal Point of Life

The centerpiece of the Asian-style garden is a swimming pond that is directly adjacent to the house and the outbuilding. When the client, climber Thomas Huber, returned from his strenuous tours, this simple single-story wooden building right by the water is a place where he can come down and regenerate. The swimming pond with its wooden walkway exudes the security of an enclosed garden space, a hortus conclusus. Thanks to a large regeneration zone, the swimming pond remains clean even without chemicals. The water lily area near the house is home to underwater plants such as milfoil and waterweed, which help to purify the water. At the pond's edge, a small strip is adorned with marsh marigolds, arrowhead, and heart-leaved pike weed. Tall reeds and irises visually separate the pond from the neighboring property.

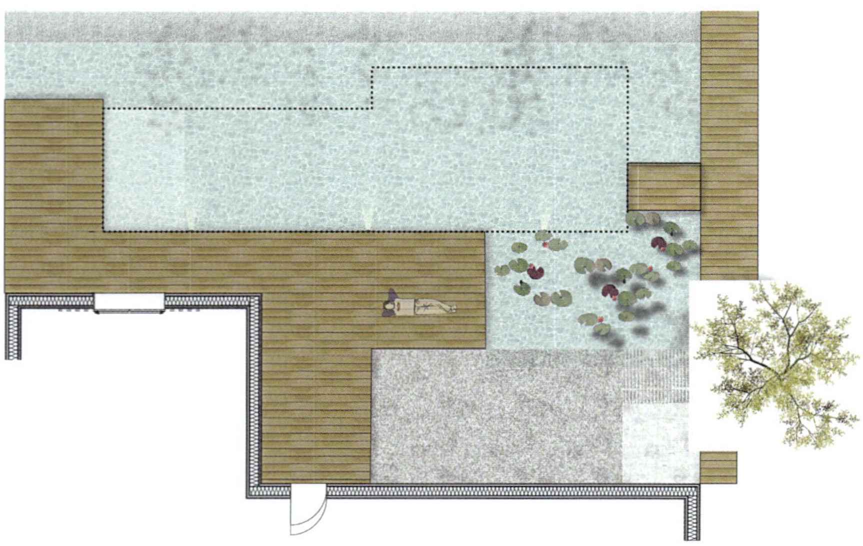

Daycare center Wolkenzwerge-Grashüpfer

On the north side of the listed Axel-Springer-Hochhaus, a daycare center for 40 children was created on a 510-square-meter area of an existing pine grove. The open space is located directly on the forecourt, that's why the outdoor area of the daycare center is enclosed by an 1.80- to 2.10-meter-high fence. The fence system has increased security requirements. A planned lamella-like construction prevents direct views into the open space of the daycare center, while the children can observe the forecourt. A playscape made of colorful panels, as well as climbing ropes, boulder buttons, and marble runs invite the children to engage in a variety of play activities. A play strip with a color gradient, which alternates between blue-green and yellow-red tones, reflects the façade of the high-rise building.

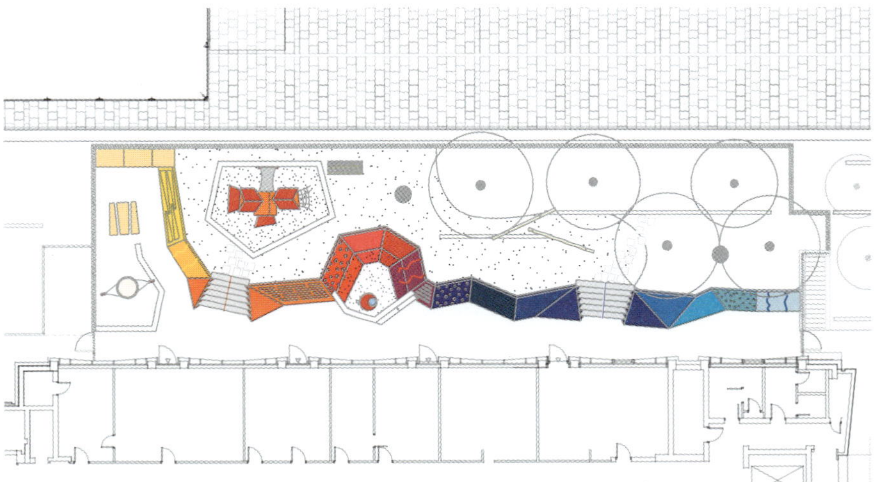

Playground ELSE

The playground combines an artistic approach with a sporting trend. Red elliptical shapes create a striking place that confidently asserts itself in a gap between buildings. The play ellipses offer innovative activities that represent a special sporting challenge. They are based on popular TV obstacle course shows such as Ninja Warrior, where contestants compete against each other in various disciplines using power, endurance, physical, and mental strength. The playground is located on Else-Lasker-Schüler-Straße. This street was named after the avant-garde poet and illustrator (1869–1945) who lived not far from the site. Inspired by her poetry, a modern abstract fantasy world was created. Designed to challenge the creativity of the individual, it leaves each visitor free to invent new dream worlds for themselves.

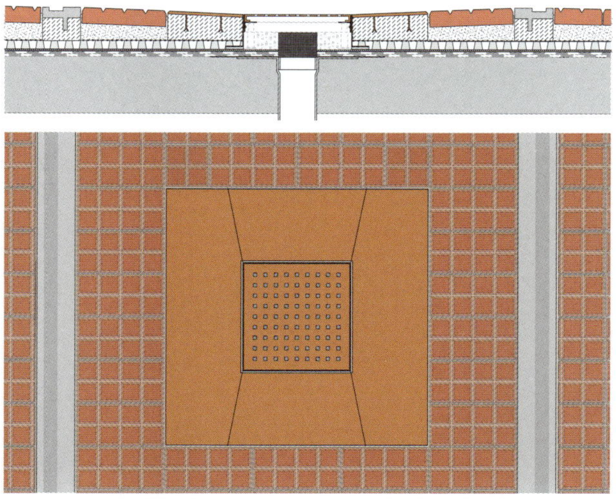

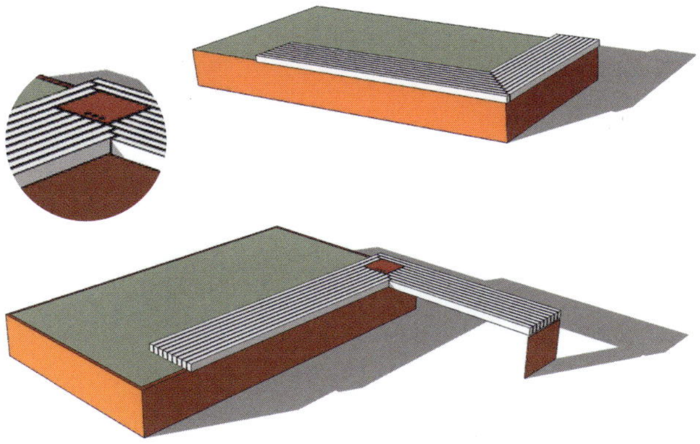

Umspannwerk Humboldt

Built in the 1920s by Hans Heinrich Müller, the Humboldt transformer station is one of the most important industrial landmarks in Berlin's Prenzlauer Berg district. Preserving its original character was a priority when designing the outdoor facilities. This involved adding furniture and creating new recreational areas. Seating and planting beds blend into the geometry of the building layout in front of the main structure, while the use of rusting steel pays homage to its industrial history. To restore the courtyard pavement surrounding the historic control room, bricks were replicated using traditional methods and combined with the original stones. The surface drainage had to be redesigned and new courtyard drains were installed. The transport tracks, originally used for transformer maintenance, were relocated to their historic positions.

Gleislinse – Former Shunting Yard in Berlin-Schöneweide

The project is transforming a former railroad property while creating new open spaces for residents and workers. During the realization of the new Gleislinse, an 800-meter-elongated green corridor, special emphasis is placed on nature and species protection. The design is based on the existing track and vegetation structures of dry grassland and ruderal areas. The remains of the former use were not only preserved, but also integrated in the design: Stacked concrete railroad sleepers serve as rhythmic, structuring landmarks, while benches made from wooden sleepers invite visitors to linger. Information boards along the way provide information about the history and ecological diversity of the site.

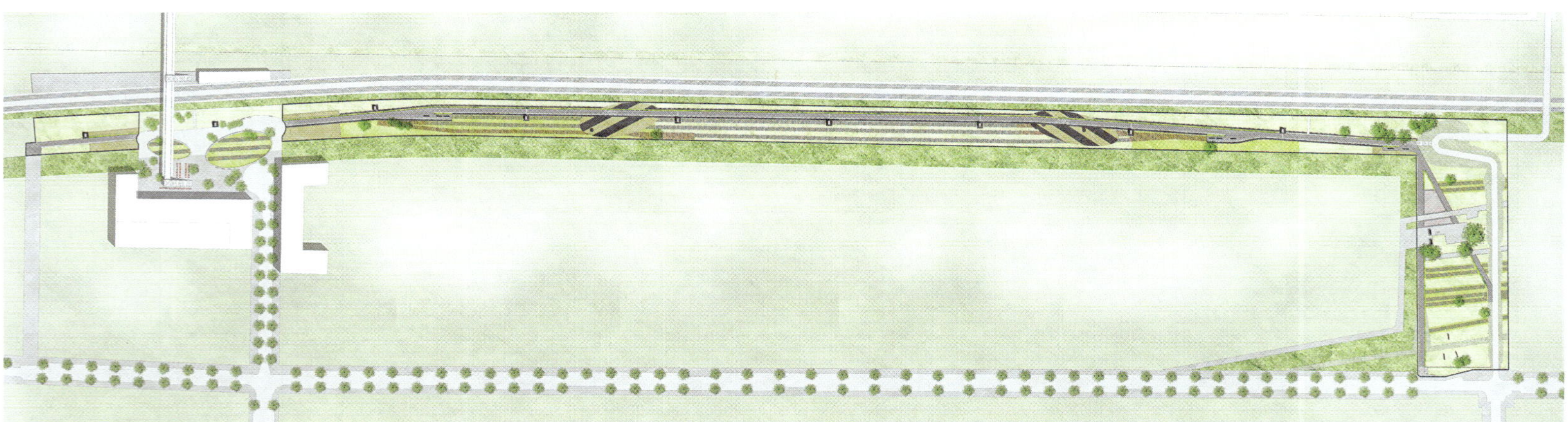

02
GLEISLINSE

African Cabinet Garden

The International Garden Exhibition Berlin 2017 invited landscape architects from around the world to design cabinet gardens. The resulting Gardens of the World are now permanent features of the Berlin park. Greeninc was the only firm invited from the African continent, and the African Cabinet Garden is based on the concept of landscape's ability to transcend man-made boundaries. It features plants from Africa that are widely cultivated in Europe, displayed in pots on a steel framework reminiscent of a skeletal ship's hull. The stainless steel and copper parts of the pots were made in Germany, while their bases were hand-woven by rural artisans in KwaZulu-Natal, South Africa. The ground plane of the garden mimics a sandy shore, where the ship appears to have beached, and is planted with native European plants.

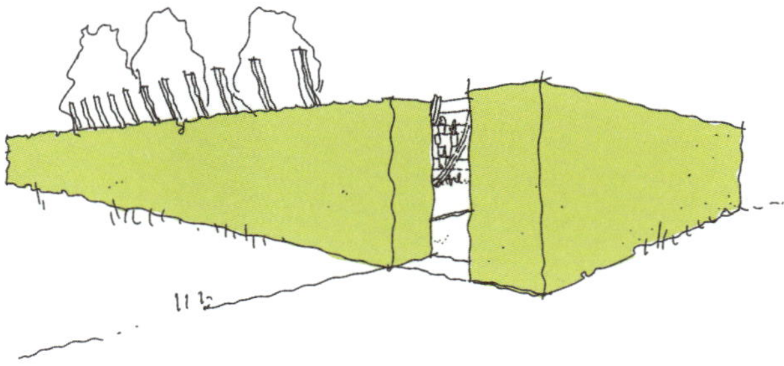

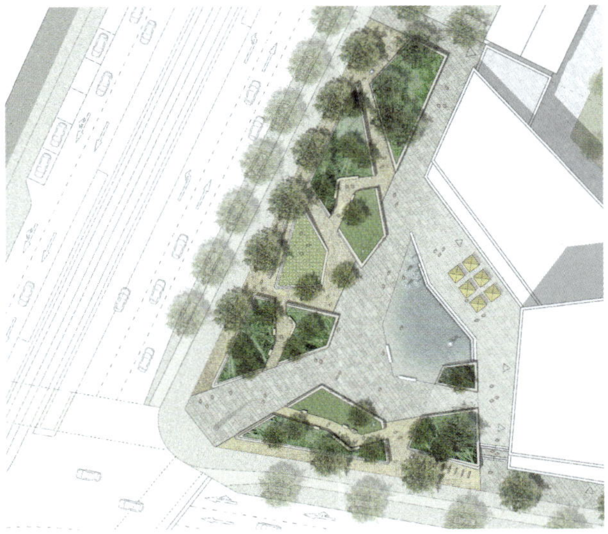

Stefan-Heym-Square, Berlin

The development of the area, including the adjacent residential quarter and a high-rise office and residential building, created new requirements for the Stefan-Heym-Square. The purpose of the project was to design a cross-generational and cross-milieu open space that meets the various demands of the surrounding neighborhood. A green framework accommodates different functions and infrastructures, shielding the square from the surrounding street. Small trees and the planting of shrubs and grasses provide light shade and recreational spaces. In between, a small play and exercise area was created in the same formal language. A shallow water basin offers cooling, while the rippling water feature fills the space. The square offers room for a wide variety of events, while the water basin can be drained to extend the area.

Anchor Gardens in Bielefeld

The project developed a modern residential quarter on a former industrial site of the Ankerwerke in Bielefeld, which used to manufacture sewing machines and cash registers. By dismantling a constricting transverse bar of buildings, the existing courtyard was extended to 3,000 square meters and redesigned as a modern hortus conclusus, creating the idea of anchor gardens. The design is based on a zig-zag path that serves as a barrier-free route and references the history of the site, as it resembles stitch patterns of a sewing machine. The anchor gardens are located on the north-facing slope of a forest and the courtyard has a steep gradient of three meters. In the center, the path widens into a square, whose red plastic surfacing forms a contrast to the up to 1.70 meter high grass waves, that make the path disappear into the lawns.

Redesign Jahnplatz

Jahnplatz is the intersection of several important and formerly busy inner-city traffic arteries. With its light rail stop and numerous bus stops, it forms an important junction in Bielefeld's public transportation system. As part of the redesign, the traffic areas were restructured and prioritized, the quality of stay was increased, and vegetation areas were implemented. To this end, the paths were combined by shading the paving. Attractive places to linger were created in the form of wooden urban sofas, and the central Alcina clock was fitted with a round bench so that it can now function as a new meeting point. Street trees to the west of Jahnplatz and along Niederwall define the contours of the square.

Bus Stop Canopies Jahnplatz

The redesign of Jahnplatz, where the light rail and all buses serving the city center intersect, was an important contribution to climate protection and improving the quality of life in the city. In addition to reducing the number of lanes for private motorized traffic, the aim was to provide light rail and bus passengers with a weather-protected boarding and transfer area. The new canopies consist of groups of umbrella-like steel structures. Luminaires concealed within the roof structure cast the light outwards through the perforated sub-cladding of the roofs and provide a uniform light pattern to illuminate the covered bus stop areas. The characteristic shape and appearance with irregular points of light give the square a new identity. Green roofs increase biodiversity in this urban area.

Board Game

One of Bochum's largest and most important inner-city squares, the Husemannplatz, will be completely redesigned in the coming years. The city of Bochum saw the last phase before the renovation as an opportunity to try out the inner city as an open space and to make playing in the city center a tangible experience with unusual ideas. The board game breaks the rigid symmetry of the square. About 90 boards are laid across the square like a network of narrow webs. The boards run horizontally, dip under each other, are sloped, or leave gaps to be overcome. It is a place that is not a playground but a sculptural, playable landscape open to everyone, not hiding its temporary and experimental character.

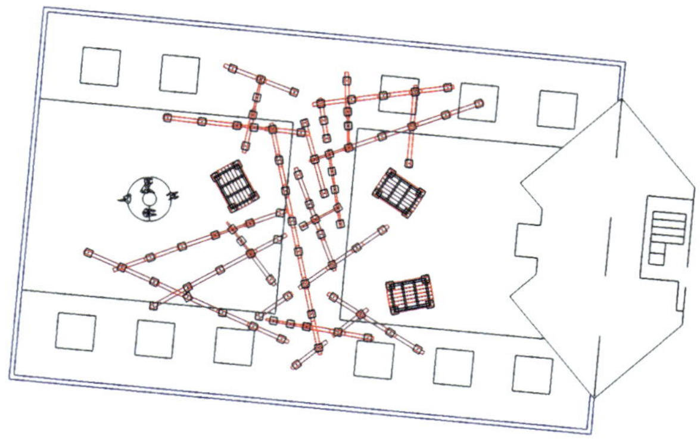

Mönblick House Garden on the Island of Rügen

The thatched half-timbered Mönblick House near the Baltic Sea and the island of Hiddensee sits at the center of its garden. It is enclosed by a frieze wall of moraine stones planted with roses and lavender. Topiaries, large shrubs, numerous grasses, and perennial flowering plant species are coordinated in terms of color, season and design. The walkway to the house was paved with natural stone. The path is framed by smaller planting areas and wooden groynes. Footpaths and driveways were edged with a five-millimeter-wide and at least 150-millimeter-high steel strip made of weathering steel. This type of edging provides a contrast to the natural stones. Architectural hedge elements were planted to demarcate it from the adjoining neighboring properties, although no important visual connections were omitted.

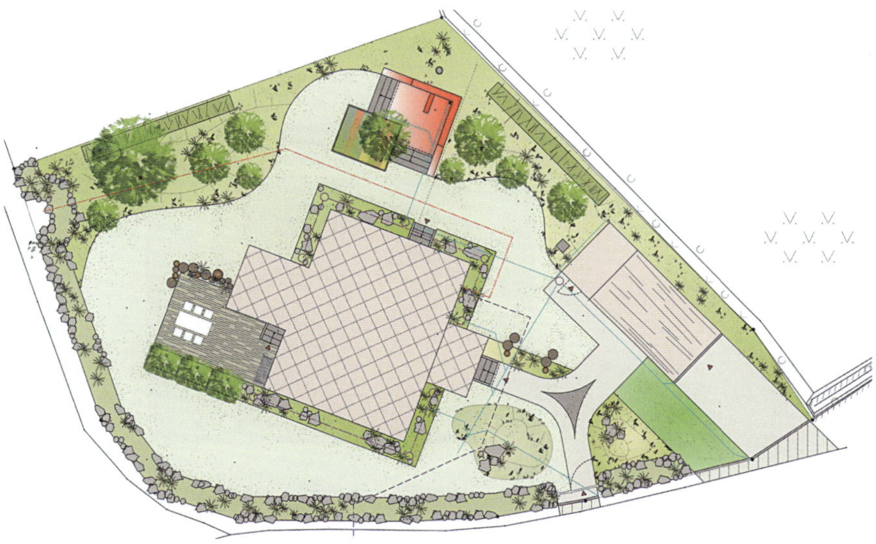

Park Planning and Outdoor Facilities at MHK Campus

The holistic design concept for the MHK Campus combines interior and exterior space in terms of function and design. The semi-public park takes up the principles of classic park scapes and interprets them in a modern way. Visitors and employees can stroll around, relax, but also work. The main boulevard, with the magnolia grove as a central link, connects the formally designed entrance area around the Headquarters with a large water basin and the cultural forum with the soft forms of the blossoming park landscape. Differentiated plantings under a varied large tree canopy alternate with open meadow areas. Sustainable rainwater management and ecologically valuable areas are integrated.

Residence at Postplatz

The project converts the old existing building of the former head post office into a residential and commercial space. The aim was to create an attractive inner courtyard and memorizable address. The task of the open space planning was to align different usage requirements. A grove and shrub structure was successfully established in the overheated inner city areas. The large areas created, planted with perennials and loose shrubs, turn the inner courtyards into green oases of relaxation and establish a green jungle in the middle of the city. Access areas, private garden spaces, an outdoor terrace, and versatile recreational areas were build for residents and public. The diversely structured shrub and perennial planting serve as a flowing transitional element between private and public spaces.

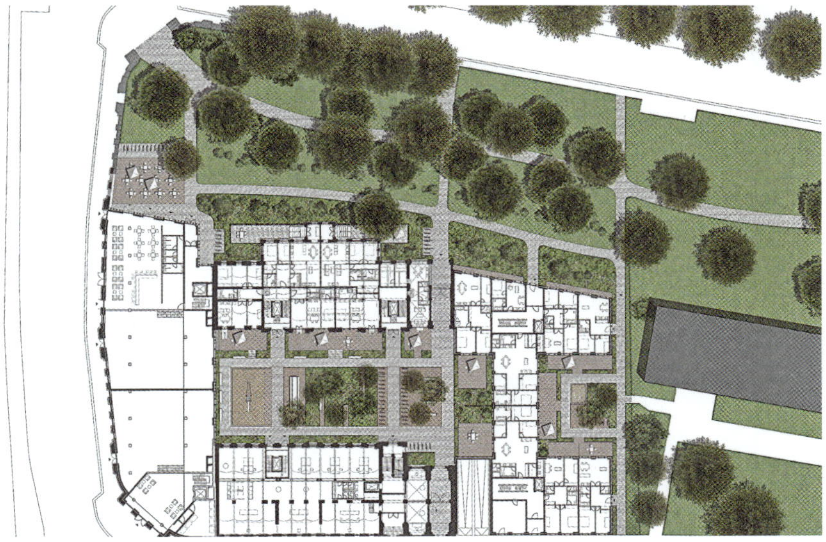

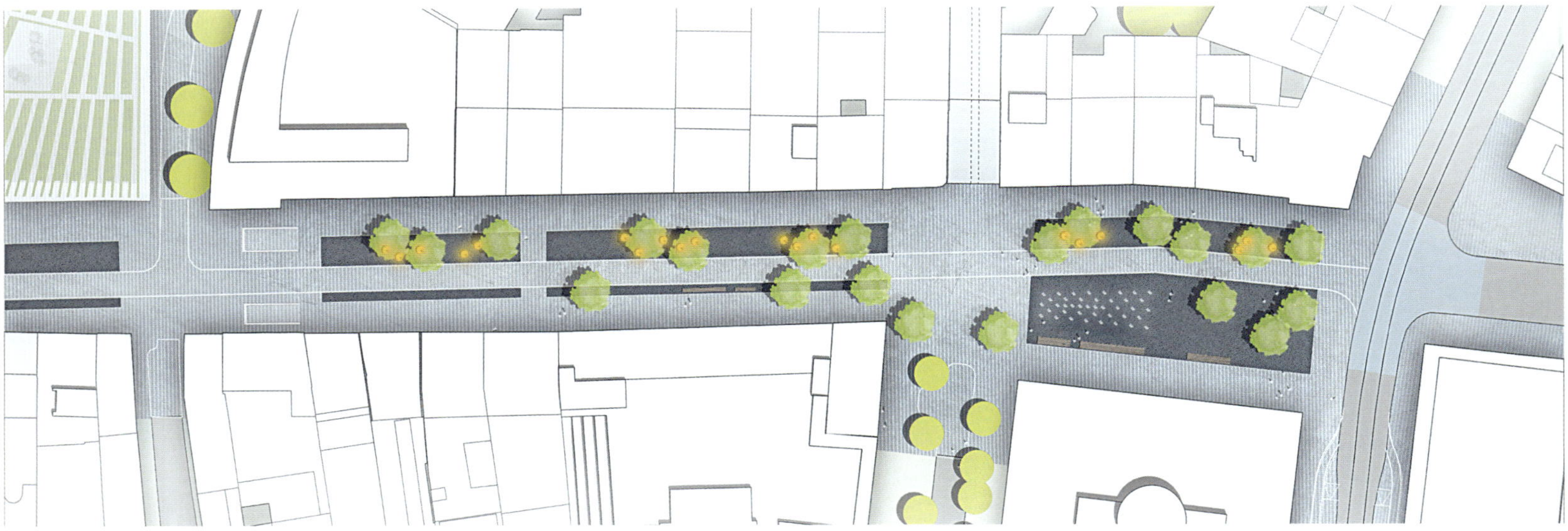

Redesign of the Pedestrian Zone Schadowstraße

Where tramways and cars used to run a few years ago, pedestrians and cyclists now bustle about. Schadowstraße, one of the busiest shopping streets in Germany, has been given a completely new look thanks to consistent traffic calming and the underground routing of public transport: Bands of terrazzo asphalt in a light-colored pavement structure the scene, while gleditsias and alders seem to dance through the street space. The bright yellow seats are particularly eye-catching and invite people to linger, while children have fun playing in the fountain. The lounge-chairs, called Shadow, are concentrated on one of the strips. On hot summer days, mist jets provide pleasant cooling.

→ kerck+partner landschaftsarchitekten → Memorial and Ecclesiastical → 2021 → Emmerthal → Germany → Photos: Volkmar Kerck

Bückeberg Documentation and Learning Site

Every year between 1933 and 1937, the so-called Reichserntedankfest, a harvest festival, took place on the Bückeberg near Hamelin. With up to one million participants it was one of the largest mass events of the Nazi regime. The new historical-topographical information system makes it possible to decipher this architectural and cultural monument, which still stands out today. The site is made accessible via a network of paths. Visitors are invited to search for clues at this extracurricular place of learning and, by walking across the former fairground, gradually gain information for an examination of the history of the site. Lawn paths within the slope, which is also used as a sheep pasture, as well as robust concrete paths and footbridges over the remains of the foundations of the grandstand connect the information islands.

Eppingen Garden Show

The Eppingen Garden Show has created a contemporary park that is characterized by a wide variety of designs and uses. The open space close to the town center combines green and water areas in a convincing way and integrates them pleasingly along the historic half-timbered town. In addition to a continuous green park strip, long-awaited access to water has been realized. Extensive renaturation measures have restored the lost connection between the Elsenz and Hilsbach streams and their floodplains, making it possible to experience the waters. Permanent structures and elements, such as the ribbon of fountains along the former mill canal and the town pond, provide positive impulses for the quality of life, the social environment, and the urban climate of Eppingen.

terroir f – Twenty Magical Places of Franconian Wine

In the language of wine, "terroir" refers to the natural environment in which a wine is produced. The terroir f project consists of 20 different sites in the Franconian wine region that allow visitors to experience the wine, landscape, views, culture, and the overall complexity of the region in a special way. Over the course of ten years, a team of landscape designers, urban planners, and graphic designers developed and implemented unique solutions for each site. Existing structures such as vineyard cottages, viewing platforms, and walls were incorporated, renovated when necessary, and complemented with modern landscape architecture. The result is an immersive exploration of the region, where every site is unique and never the same.

Main Riverbank in Frankfurt by the Ruhrort Shipyard

In the final section of the transformation of Frankfurt's eastern harbor into a post-industrial urban landscape, which was difficult to equip, a varied green space was created on the former Ruhrort shipyard. The result is a differentiated system of paths for pedestrians and cyclists. It emerges from the intersection of two superordinate green routes: the Main riverbank promenade and the Frankfurt green belt. It is an extensive network of flowing park paths and meadow areas with constantly changing flowering aspects. The promenade on the banks of the Main is equipped with robust wooden platforms and seating furniture. The memorial to the deported Jews and the relics from Frankfurt's industrial past keep the history of the site alive and accessible.

Gehrden City Center

The heart of the new 500-meter-long design of Gehrden's central inner-city shopping street is the market square with the adjacent churchyard. The churchyard, located one meter above the market square, now leads into the market square via an ellipsoidal staircase with seating and a ramp with new pathways and play equipment. In the middle of the steps is a water feature – designed by Gansloser engineering office – the water of which flows over the steps. The square itself and the churchyard have been redesigned with terracotta-colored clinker bricks. The façades and the church are illuminated to create a striking spatial effect. The lighting designers were SchmitzSchiminskiNolteMathee Design, while PVB-Planungsgemeinschaft VerkehrsBau was responsible for the civil engineering.

The Johannisberg Monastery

Kloster Johannisberg, a former Benedictine monastery in Rheingau, was acquired by Claus Wisser in order to preserve its past, refill it with life, and transform the monastery grounds into a radiant cultural center with a newly created park. The result is a contemplative but classical paradise garden in a contemporary interpretation. The cloister garden is surrounded on all sides by a trellis of olive willows. Three solitary apple trees, arranged in an uneven triangle, provide an harmonious counterpoint to the clear structure of the inner courtyard. The symbolism of the cross not only permeates the spirituality of the place, but is also visible in the paving stones, which create a cross mosaic. The color of the dolomite forms a pleasant contrast to the lush greenery. A fountain field provides a cooling atmosphere in the summer months.

Multi-Generation Park

Located in the heart of Goldenstedt town, adjacent to a catholic kindergarten, a nursing home, and a multi-generational complex, the park aims to provide a natural gathering place for people of all ages. The park's design focuses on three main themes – culture and communication, physical and mental activities, and experience and recognition of nature. A stage was installed as the centerpiece for the first theme, hosting music events in the summer. The so-called Family-band area offers an array of mental and physical activities such as street basketball, senior fitness equipment, table tennis, a play space for toddlers, a climbing structure for kids and teens, as well as bowling and chess fields. Within this area, the community can participate in gardening activities at the orchard, encouraging visitors to experience and learn about nature.

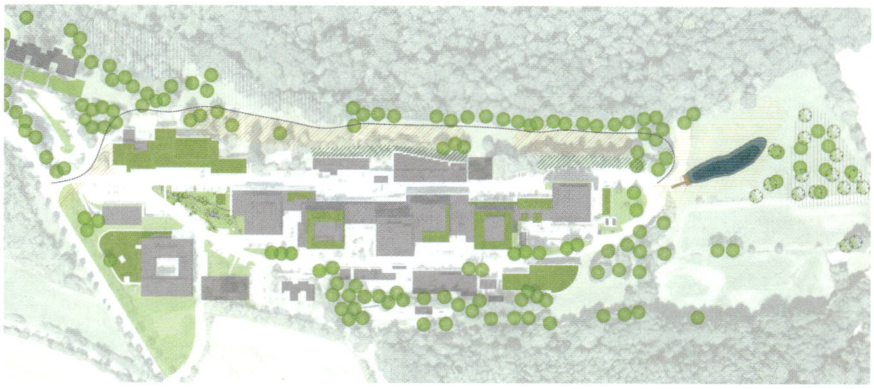

BioDiversum

On the initiative of the ornithologist professor Peter Berthold, a plan to reshape the grounds of the Max-Planck-Institute for Multidisciplinary Natural Sciences was made in 2019. The redesign was made in order to counteract the sharp decline in animal and plant species. With the technical support and coordination of the Institute, the studio Wette+Küneke Landschaftsarchitektur created the BioDiversum, a pilot project within the Max-Planck-Society. Since the renaturation, the living conditions for insects and birds, but also for many other animals, and plants have already improved noticeably. This has been proven by scientific studies. During their breaks, employees benefit from a natural working environment with pleasant places to spend time. The entire campus is open to the public and information boards have been set up.

Planten un Blomen (Plants and Flowers)

As part of the revitalization of the Congress Center Hamburg (CCH), the adjacent Planten un Blomen park was extended and redesigned along with its surroundings. Among other things, the four-lane Marseiller Straße was dismantled and filled in. Now there is a sequence of plantings and squares, including a park balcony with a view over the redesigned Dag Hammarskjöld Square. A barrier-free path leads from the historic station portal across the former forecourt of the former Dammtor Imperial Railway Station along an intensively planted botanical showcase, with a variety of staged views of the park, across Stephansplatz and directly into one of Hamburg's most beautiful parks. Dag Hammarskjöld Square has thus been transformed from an unattractive rear side of the station into the new antechamber of Planten un Blomen.

Ada-Lessing-Park

In a quiet, green residential area, a multifunctional park was developed on the fully sealed areas of the former Linsingen barracks. Five anchor points with large seating elements open up the park to its surroundings. The curved pathways between them also serve as superordinate cycle path connections. Four green park areas, each with its own character, allow for different uses: The free meadow offers plenty of multifunctional open space. The climate forest provides an overview, views, and shade under a climate-friendly canopy on a gentle hill. The moist sink serves as a retention area for the planned large vocational school complex next door. The urban arena offers modern leisure sports and exercise facilities.

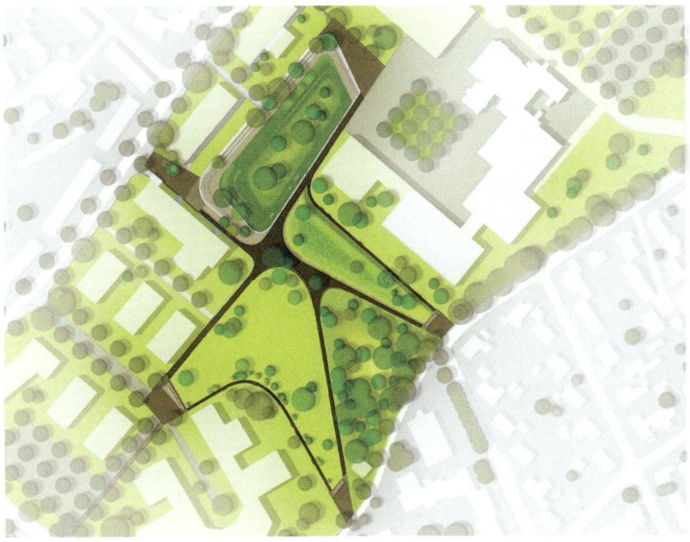

Summer Island

The Summer Island was created as part of the Federal Garden Show in Heilbronn as one of the first completely 3D-generated and built landscape parks in the world. On an area of around 4,200 square meters, organically shaped waves of landscape were laid together, combining the natural forms of dunes, moraines, and the ephemeral patterns of a sandy landscape. The BIM-based planning of the topography enabled rapid progression in variants, real-time mass and cost control, and a short construction time thanks to the effective interlocking of the interfaces. The result is a landscape that can be experienced as a walk-in sculpture and that offers a captivatingly built staging, just like the Baroque or later the English garden did. But Summer Island is a stage set for the democratic era.

Tongrubengarten

The primary objective of this project was to integrate two adjacent plots into a garden that invites the visitors to take a stroll. The paths that wind through the vegetation are a recurring theme in the design made by Maria Hänsch. Narrow, curved paths lead through the extensive garden to various vantage points, such as the seating area by the canal at the edge of the property. This makes an ideal visual axis. Standing at the garden gate, an unusual privacy screen made of peeled robinia wood ensures seclusion. With its natural design, however, there is nothing unapproachable about it; in fact, it is an eye-catcher itself. What makes the garden particularly interesting is the great diversity of species. Robust plants and grasses were necessary due to the site being a former clay mine for brick production.

Kindergarten Schlehenstein

The Schlehenstein municipal kindergarten is housed in a villa from the 1920s. The large open spaces have been extensively renovated, retaining the generous character of the outdoor area with its existing trees and shrubs. From the entrance area with its reused natural stone paving, the garden can be accessed via a grassy embankment. Here is a central recreation and play area: sand pits, bordered by wooden decks, arranged at different heights. In the southern part of the garden, areas for play equipment have been laid out in a polygonal shape, delimited by play paths made of in-situ concrete that form a driving course. The so-called painting house is also located here. The property is bordered by a wooden fence, with different slat widths on a concrete base, and a hornbeam hedge.

Education and Sports Campus Kreuztal

In the course of an ideas workshop, the aim was to develop a slope between schools and a sports hall into an attractive campus. The studio wbp landscape architects designed a concept with a center open for versatile use, around which various intergenerational facilities are located. The result is a multifunctional arena designed with grass steps in the center of the campus. The existing leisure facilities such as the skate bike park and the small playing field were upgraded and expanded in close consultation with the young users and supplemented with a play slope and a calisthenics facility.

Lakeside Park in Lahr

The Lakeside Park came into existence in 2018, when the State Garden Show initiated a rejuvenation of brownfield sites in the city of Lahr. Characterized by a newly constructed lake, extensive meadows, and clusters of trees at the periphery, the park not only offers a variety of recreational opportunities, but also creates a habitat for regional flora and fauna. The lake promenade lined with willow trees and the lakeside terrace, which is distinguished by a cafeteria and pine groves, form the active core of the park. A riparian forest on the south side invites people to walk through the herbaceous understory and discover the small fern islands. The viewing platforms and wooden bridge offer a breathtaking view of the city across the lake.

Residential Courtyard with Small Marina at White Elster

The residential courtyard was built alongside an apartment building on the grounds of the former Buntgarnwerke Leipzig dye works. The location of the Weiße Elster river banks was pivotal to the courtyard's design, ensuring all residents can directly experience its advantages. Therefore, an amphitheater-style marina was created, completed with spacious green steps that encourage residents to unwind and enjoy the view. Featuring to a floating dock, the marina provides a docking space for small recreational boats. The historical bank wall has been reconstructed. In the upper garden area, children can enjoy playing with monkey bars, a sandbox, and large river pebbles. High-quality materials such as granite pavers, natural stone retaining walls, and selected trees and shrubs have been used to ensure durability.

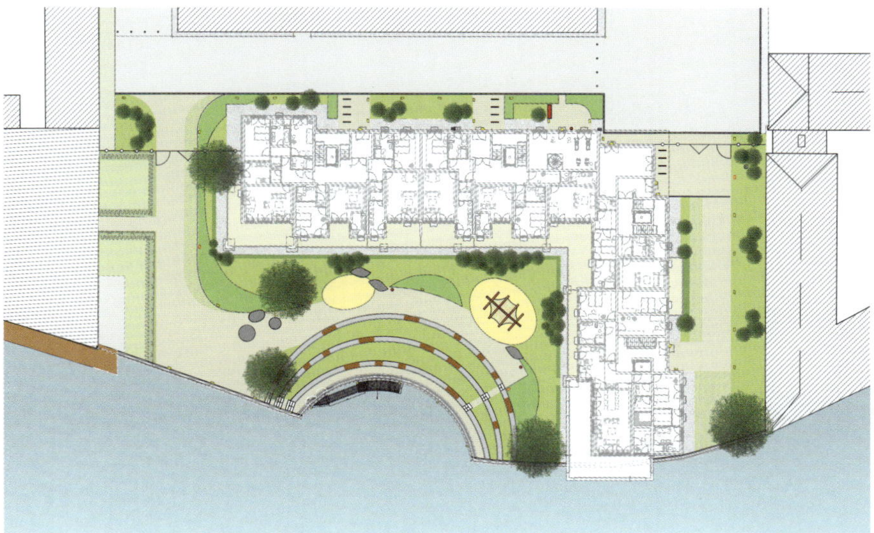

Campus FH Applied Sciences

The outdoor facilities of the new campus for the applied sciences in Leverkusen at Cologne University of Applied Sciences were realized as part of the development of the New Railroad City of Opladen. The design of the campus park picks up on the style of the former tracks of the railroad repair works. Strips of shrubs, grasses, and trees flank the paths and structure the open space. The interwoven strips of track ballast and the brick-paved paths are a reference to the materials that once were predominant here. Urban, barrier-free spaces, such as the campus boulevard, offer various places to spend time and ensure smooth access. At the same time, a parking facility for 150 cars, 30 motorcycles and 120 bicycles was created.

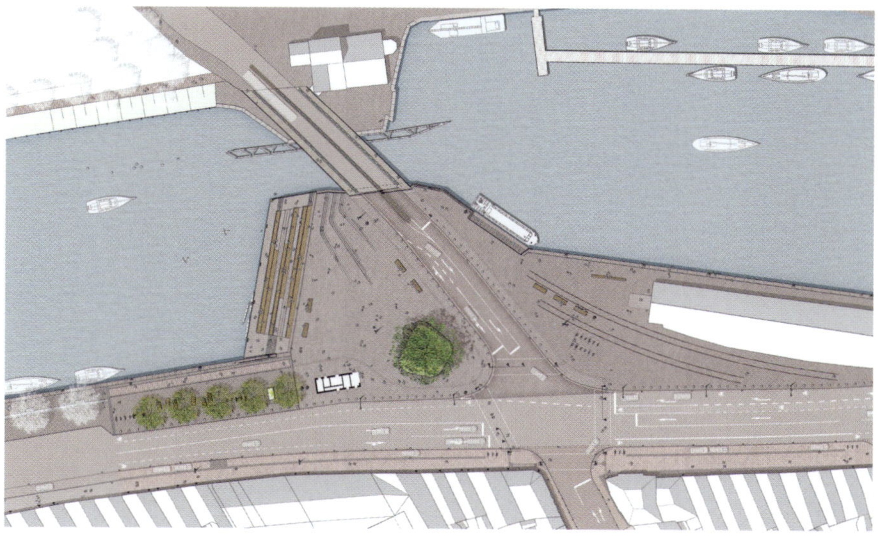

Drehbrückenplatz

Situated in the harbour, on the western edge of Lübeck's Unesco World Heritage Old Town, the Drehbrückenplatz (Swing-bridge square) stages its imposing surroundings of bridges, cranes, harbour sheds and merchants' houses by means of a broad open space. The square, featuring the distinctive stone paving typical of harbours, is punctuated solely by a large poplar tree with an enclosing bench and extends down to the waterfront via a wide flight of steps. Facing south, four seating steps offer a view of the Untertrave with its bustling quay side over a length of 40 meters. As a prelude to the new promenade An der Untertrave with trees, loungers and benches, a small architecture was created to accommodate gastronomic facilities.

Walckerpark Ludwigsburg

The Walckerpark forms the northern entrance to the city center. As part of the redesign, places of encounter and identity were created that have a high ecological value, improve the urban climate, and are resilient to user pressure and climate change. To achieve this, 150 trees were preserved and 135 additional, mainly native, climate-resilient trees and large shrubs were planted. A newly developed water-permeable in-situ concrete made it possible to completely unseal the asphalt surfaces, so that all the surface water can remain in the park. The park deck façade was completely greened and the embankments facing the street were planted with species-rich seed. Artistically designed furniture elements made of steel and wood, a golden cage for soccer and streetball as well as a playground offer opportunities for sport, play and relaxation.

Prinz Eugen Park – Ecological Model Estate

The two quarters of the ecological model estate in Prinz-Eugen-Park are highly in tune with the spirit of our times: the careful use of energy and resources and a radical restraint in the choice of materials form the basis of the planning. The designers are no longer the focus, but rather the planners have entered into an intensive, continuous planning and working process with the residents and have awakened their sense of soil, plants, and forms. The result is a garden for the residents of the neighborhood, who now gather horticultural knowledge, exchange ideas, and nurture and care for the open space together. The ecological model settlement was realized through the commissioning by the City of Munich, which stipulates the implementation of ecological purposes.

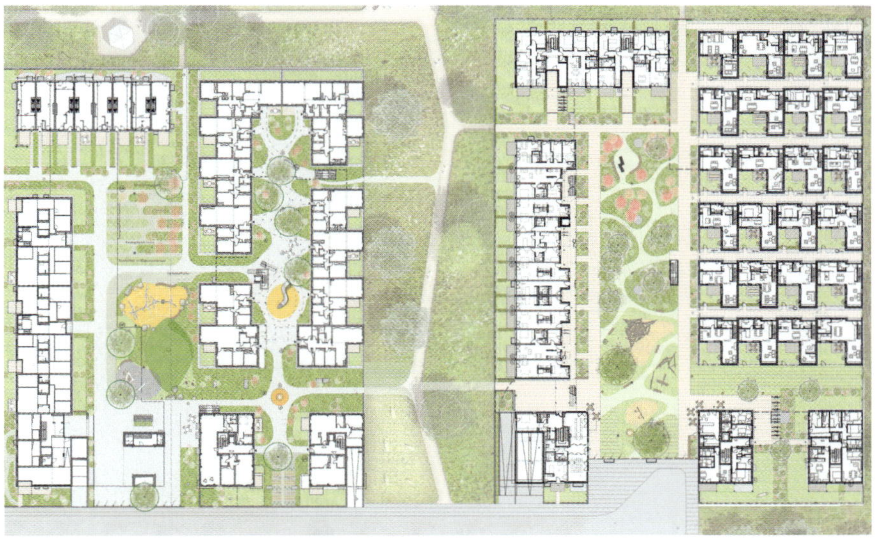

Station Gate Neubrandenburg

The 8,000-square-meter open space design between the train station and the old town makes the city's history recognizable and perceptible. With the double ramparts, the city wall, the Franciscan monastery, which is now used as a regional museum, and St. Johannis Church, there are important garden and building monuments that have shaped Neubrandenburg since the Middle Ages. Stelae, which reproduce the cross-section of the medieval wall profile, trace the course of the town fortifications at different intervals and condense towards the street to form a new gateway situation without closing off access to the town again. The patina of the Corten steel stelae corresponds with the brick of the adjacent listed buildings. Plaques with information on the town's history were incorporated into the stelae.

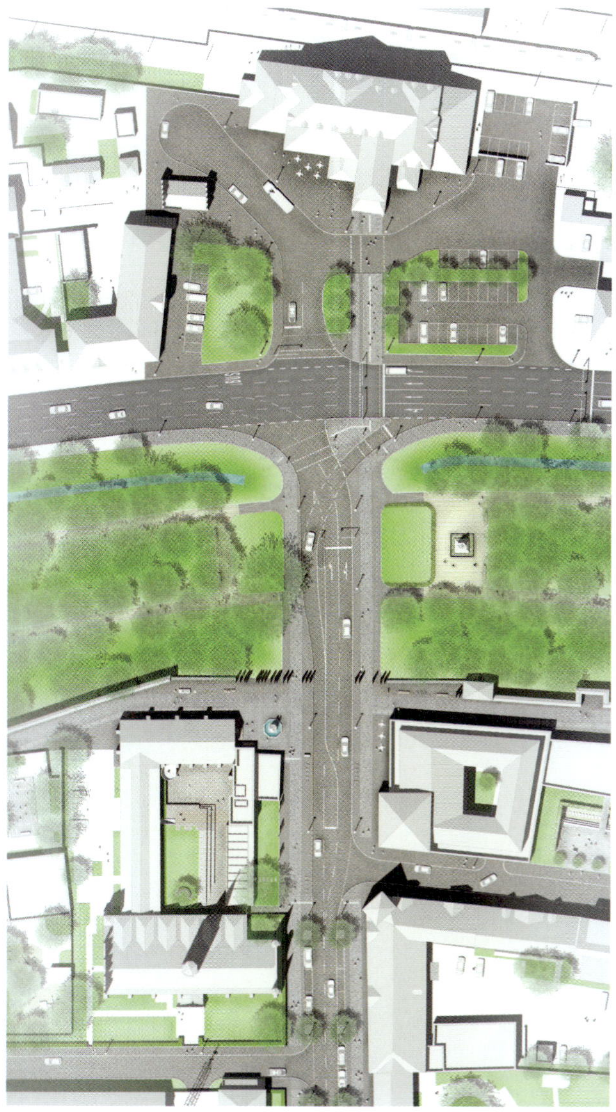

Serenity Park

The park section of the over 10,000-square-meter property is stylistically reminiscent of classic English landscape gardens. There is a central lawn, curved paths that are repeatedly accompanied by hedge elements, and grass borders that alternate with cherry laurel and rhododendron borders. To the south of the lawn is the Axis of Serenity. This is designed with a slight slope of concrete slabs in the lawn and leads into a Japanese garden with a large koi pond, the edge of which is secured with boulders and a round-washed river stone. Japanese elements such as stone lanterns and small pagodas accompany the banks of the pond to the teahouse with terrace. Rhododendrons, Japanese fan maple trees and sequoias create a sheltered character and lend the Japanese garden a meditative atmosphere.

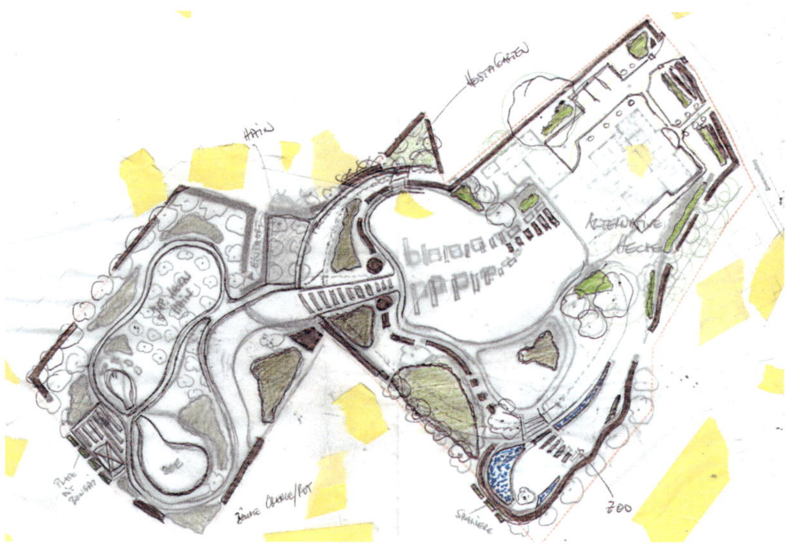

University and Science Campus Westerberg

The university campus in Osnabrück is gaining trans-regional importance and has been expanded and developed to include the 55,000-square-meter military site of the former Von-Stein-Barracks. The existing site potential is being used in dialog and synergy with the extensive topographically attractive landscape to the south. The ensemble around the former state engineering school, built in 1966, is characterized by generously designed open spaces with differentiated levels, terraces, staircases, and clusters of solitary trees. The new urban design complements the pedagogical and architectural ensemble and creates new options for students, teachers, guests, and citizens. Integrated into the city's educational landscape, the new campus offers a sociocultural urban laboratory for communication, culture, art, and science.

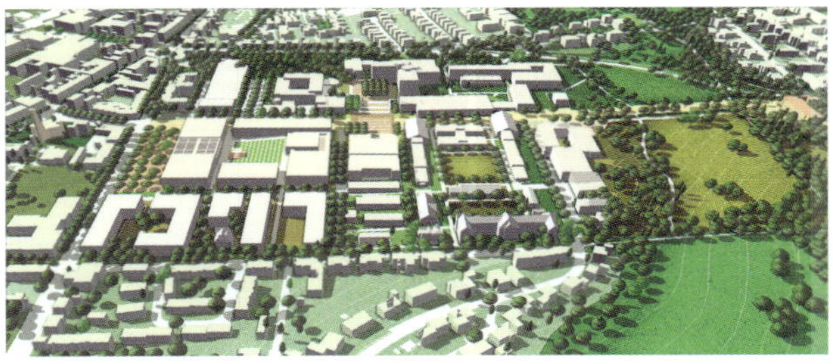

Alfred-Wegener-Institut – Office and Laboratory

The outdoor areas of the Alfred-Wegener-Institut blend the building with its environment by using indigenous trees and shrubs that harmonize with the local topography. The design echoes the themes of the Institute's work, which focuses on the exploration of polar regions and oceans. The central theme of the forecourt focuses on expeditions to the Arctic Ocean, while the courtyards symbolize the permafrost soils of the tundra. The forecourt is structured by paths and lawns and leads up to several generous building entrances. The lawn areas are interrupted by white polygonal slabs, which symbolize ice floes and are inscribed with keywords indicating research sites. Linear steel walls emphasize the cascading steps that structure the forecourt, and embedded granite slabs evoke the glacial landscapes of Greenland.

Schopfheim School Campus

The renovation of the Johann-Peter-Hebel School along with a new building for the Friedrich-Ebert School created a diverse school campus with an optimal learning environment for the students. At the center of the schoolyard lies a wooden deck, punctuated by rock pears and surrounded by a robust asphalt surface. All circulation areas are paved to match the color scheme of the architecture. However, the school grounds offer many opportunities to discover and learn about nature, too. The planting is biodiverse and varied. Selected plants require minimal maintenance and are resistant to environmental influences. Gabions and concrete blocks create areas where pupils can relax and retreat. The play area next to the new sports hall offers a variety of adventures such as boulder walls, climbing equipment, and willow plantations.

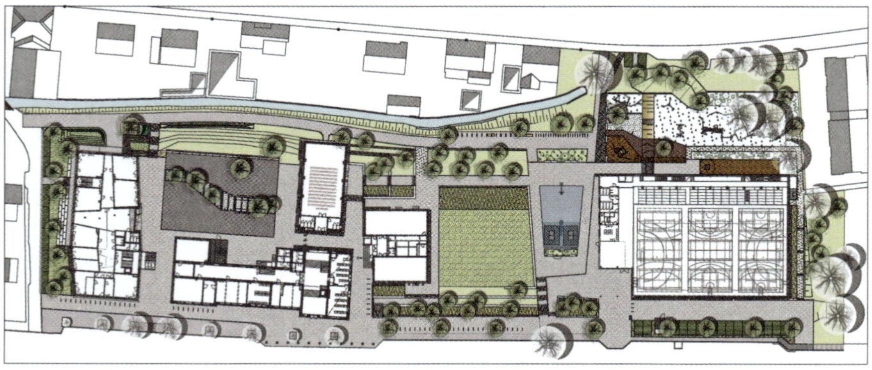

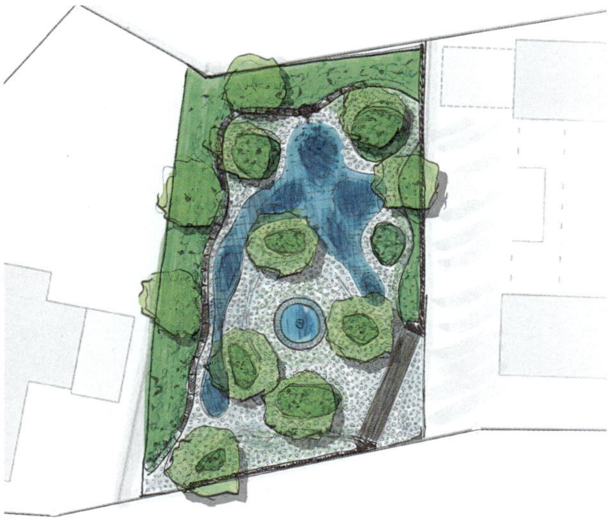

New Water Garden in Einruhr

The Water Garden is located next to the Eifel National Park in in Simmerath's Einruhr district and was designed as a recreational area for tourists. Situated along several bicycle and hiking trails, the garden invites visitors to sit on benches and chairs, walk through the shallow water, and enjoy the view of the Eifel landscape, while listening to the gentle sounds of the water. The aim of the design was to create a place where water could be experienced in all its diversity. A grand entrance staircase was implemented, and walls and fences were removed to open up the space and allow for unobstructed views and a calming atmosphere. Natural materials were used in the construction, such as graywacke for the stone elements, wood for the seating areas, and landscaping elements in the plantings. The water park responds to its surroundings by referencing the nearby Heilstein spring.

Hillside Garden on a former Vineyard

The Hillside Garden, a private garden in Ahr valley, is located on the steep terrain of a former vineyard. Inspired by the walls and rocks of the volcanic Eifel region, natural stone is the main element of the project. Seven levels with a height difference of around 40 meters make the garden a naturally beautiful place throughout the year. A basalt staircase leads from the rocky front garden past the house, entering an outdoor room with a vegetable garden and seating made of graywacke. Above is the water area with a waterfall and a small Kneipp basin next to a formal pavilion. The following level is all about the craftsmanship behind dry stone walls, as there is slate all around a winegrower's hut. The seventh level is a natural garden connected to the forest, combining huge basalt rocks weighing up to eight tons with perennials and trees.

Rudolph House

Rudolph House is located below a panoramic street in Stuttgart with a view over the Stuttgart valley. A footpath parallel to the road leads to the central piazza ten meters below, which is adjoined by the loft-like living space on the top floor and, opposite, the glazed front of the otherwise earth-covered and overgrown garage with its collection of vintage cars. The house and the piazza are embedded in a garden which, with its planting of cypresses, yews, rosemary willows, magnolias, fan maples, and the central sheltering catalpa, creates references to the Mediterranean and an oasis of peace and hospitality close to the city.

Bathing under old Birch Trees

The project Bathing under old Birch Trees creates a heightened seating and lounging area made of wood, which connects an integrated swimming pool, equipped with a closable roof, with the terrace area and the garden level. The pool garden is planted with white birch, lavender, multi-stemmed pines, lamp cleaner grass, and fine-stemmed Chinese reeds. A wide band of steps with integrated plant pockets, containing various plant beds and plant tubs, lead to the lower garden. An additional side staircase leads to the garden house with sauna. All the wooden decking is made of thermally treated ash wood, which does not warp, hardly swells or shrinks and comes from sustainable production. The project creates a peaceful, park-like atmosphere in the grand garden.

Talauenpark

Waiblingen's Talauenpark, which is close to the city center, was further developed in line with the times as part of the Remstal Garden Show 2019. Existing potential was strengthened for the future, new pathways were added and the park was upgraded with sports and play facilities for local recreation. An "artificial clearing" was created on a meadow area as the new heart of the floodplain. A grove of tall trees gives shape to the clearing. This shape inverts the volume of the Stihl art gallery. The public space it creates is now home to works of art and cultural events. The new skate park was designed as a flow park and is used for international competitions. The defining element of the facility is a bowl sunk up to 2.35 meters deep into the terrain.

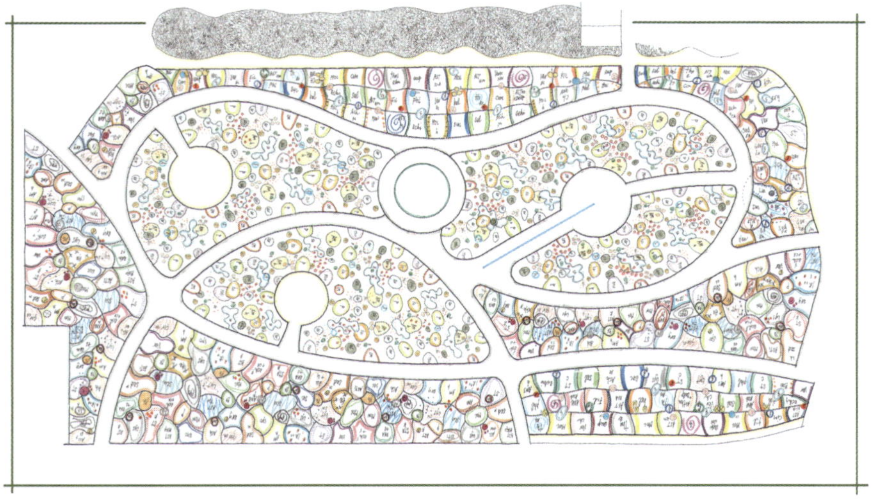

Oudolf Garten on Vitra Campus

The Vitra Campus is a historic site of the furniture manufacturer Vitra. Here, the 4,000-square-meter garden designed by Piet Oudolf blossoms in all its diversity. With his ensemble of perennials, herbs, grasses, shrubs, and meadow flowers, artfully composed down to the last detail, he has created a constantly changing wilderness between the VitraHaus and Álvaro Siza's production building, which opens up new perspectives on the surrounding architecture of the Vitra Campus. Around 30,000 plants, including plants with mysterious names such as Persicaria amplexicaule (Alba), Echinacea pallida (Hula Dancer) and Molinia (Moorhexe), form the framework of the garden. An old cherry tree and the sculpture Ring and Ruisseau by Ronan and Erwan Boureoullec are integrated into the garden.

Forecourt of St Martin's Basilica Weingarten

Weingarten Basilica is one of the most important church buildings of the High Baroque period and a popular tourist attraction. During the redesign of the forecourt, a barrier-free access and an appropriate pavement with good accessibility were realized. It serves different functions and yet appears coherent. Among other things, the area was also optimized for the fire department with a new retaining wall. The forecourt was designed with a large area of natural stone paving, which is subordinate to the basilica in terms of design. A band of natural stone slabs forms the boundary to the façade. The existing entrance platform was structurally renovated, modernized, and made barrier-free with a two-flight ramp. An easy-access guidance system and improved lighting were also created.

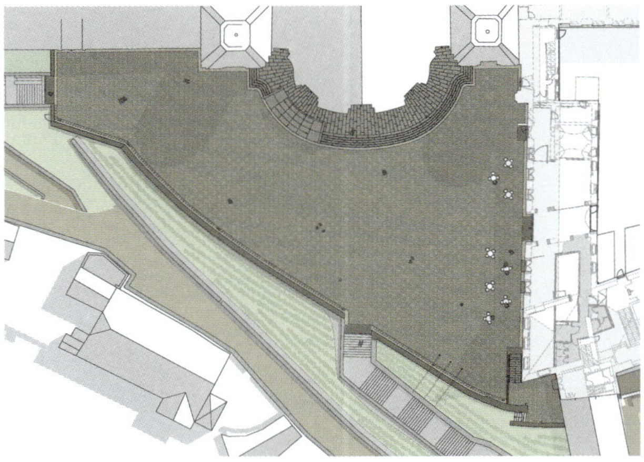

Parkstadt Wiehl

The goal of the project was to redesign the Wiehlaue, a previously neglected area along the river Wiehl, as a new urban park space in the heart of the city. The design of the this 45.000-square-meter park is centered around a small body of water. Freed from its shadowy existence under the bridge, it is now connected to the river via a new water axis. The park is divided into three areas that provide space for recreation, sports, and relaxation. A large lawn in the center can be used for a variety of purposes. A water playground, toddler area, miniature golf course, skate park, beach volleyball court, and monkey bars create a new sports and recreation space that partially runs under the bridge. To the east, a health-focused spa park with riverfront seating, water fountains, and diverse vegetation invites all ages to relax.

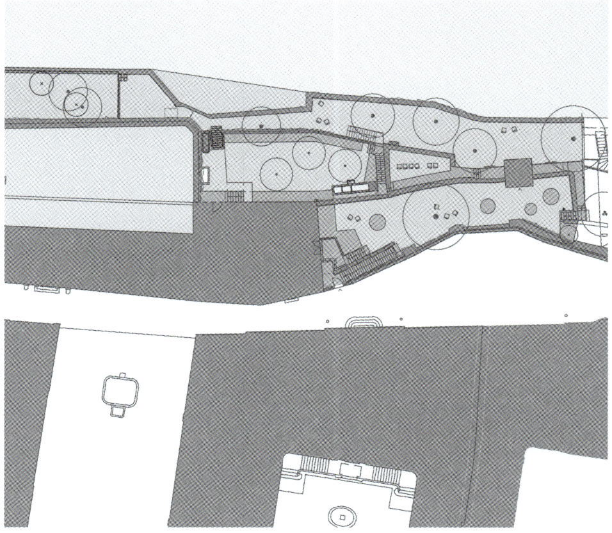

Garden of the Old University

The garden of the Old University and the public terraces are located between the Middle Bridge and the Basel Minster and fit harmoniously into the Basel city scape. The formerly overgrown garden by the Rheinsprung was renovated thanks to the financial support of the gardener Karl Schlecht. The garden has been open to the public since August 2017 and the studio bau kultur landschaft added a pavilion. The aim was to create a botanical garden, where medical students could study plants and use them for medicinal purposes. This idea came from professor Caspar Bauhin. The historic terraced garden was one of the first botanical gardens north of the Alps in 1589. It has undergone numerous transformations over the centuries. The lavishly restored and further developed garden is now a cherished place to spend time.

Single-Family House Bottmingen II

The garden of the single-family house is located in Bottmingen, a suburb of Basel, and creates a new, extended outdoor living space. Great attention was paid to the visual axes and proportions, the use of high-quality materials, the combination of colors and the play of light in the garden area. The project realized an outdoor kitchen with dining area, a lounge area with sun deck, a pool, and a specially designed atrium. Individual trees were planted for this project to provide privacy, cast shade, absorb noise from the surroundings, or to serve aesthetic purposes. Character trees in particular develop over the years and make the relationships in the garden more subtle. In the evening, sensitively coordinated lighting creates a particular feeling in the garden, which also ensures a sense of safety, security, and well-being.

Quartier Trift Horgen, Areal Kuppe

A virtually car-free, ecologically compatible and socially connected settlement is being built on the shores of Lake Zurich. The Kuppe was the first area to be completed. Five residential buildings, each with six apartments and a separate bicycle shed, are grouped around a gentle hilltop ("Kuppe" in German). A circular path around the central meadow provides access to and connects the buildings. The surrounding landscape with open meadows, fruit trees, and copses was the model for the design. A mixed planting of perennial shrubs and self-seeding annual and biennial herbs and flowers forms a dynamic and variable plant composition in the areas in front of the residential buildings. A play area, a balancing course, and a kitchen garden blend naturally into the existing topography, always with a magnificent view of the lake and the Alps.

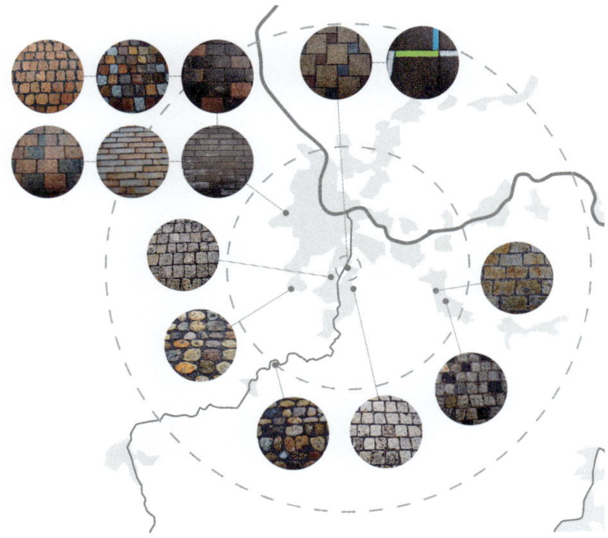

Primeo Energie

Demolition material, residual stock, and material with defects were used to create a public square on the Primeo Energie Kosmos Campus. Circular and low-CO_2 construction was seen as an obligation. The result was a square made from resources, 90 percent of which came from within a radius of ten kilometers, the remaining from within a radius of 40 kilometers. Sourcing of and constructing with these left-over materials required research into resource – dependent design and the resulting question of circular construction aesthetics. In addition to finding a design from the diverse and random availability of resources, the technical and normative hurdles such as foundations and footings had to be solved. The square is a project which asks pioneering questions that manifest themselves in the present and give confidence for the future.

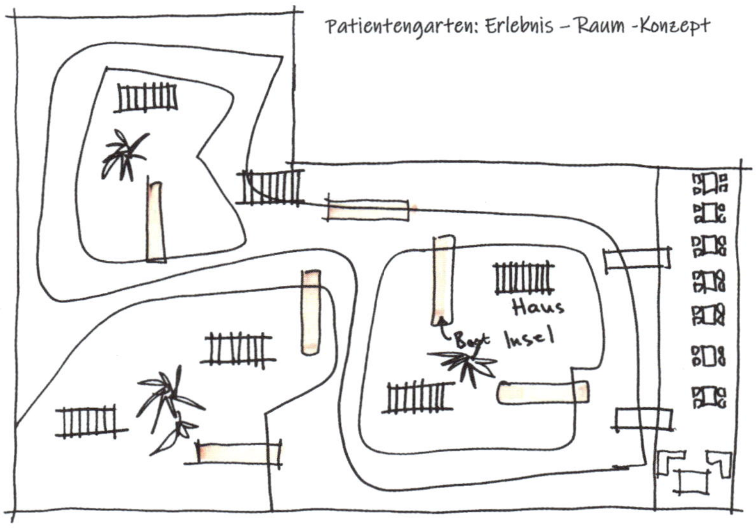

Herz-Neuro-Zentrum Bodensee Münsterlingen

Planungsgruppe Landschaft und Raum did the landscape design for the new heart-neuro-clinic building by Nixdorf Consult Gerlingen, situated on an elevated meadow plot in a rural setting with a view of Lake Constance. Fruit trees and rows of maples planted on an herb meadow mediate between the striking building and the gentle landscape stretching as far as the lake. The patient garden is located above the underground parking garage, offering impressive views on all sides. The idea of the garden is based on the design language of Oscar Niemeyer, on a curve and a straight line that creates tension and variety. Paths around raised green islands allow patients and visitors to take circular routes. Small-crowned individual trees provide shade. Rose boats lead to seating areas under pergolas, offering a view of Lake Constance.

Tournesol day nursery

The new public spaces around the Castle of Renens and the new Tournesol nursery by architect Atelier Pulver architects create a generous, inclusive, and educational place for the neighborhood. The outdoor spaces have been enhanced with the creation of a new landscaped public space that includes the development of 22 community gardens, while also integrating the more intimate spaces reserved for the daycare center. The Broye stream, which had been buried for many years, has been revealed by the new layout of the paths that connect the redesigned living spaces. Linden and wild cherry are the main trees, a number of fruit trees such as cherry, pear, apple, and plum have also been planted. A combination of shrubs act as a visual filter, and hygrophilous vegetation has been favored near the stream. Client was the Ville de Renens.

House Widacher – Garden Courtyards

Widacher is a new building with apartments and a work studio for people with disabilities. The arrangement of the three buildings gives all the residential groups an east-west orientation with their own sheltered outdoor terrace and access to a green garden courtyard. The garden courtyards allow residents to find sheltered meeting places, and relaxing retreats. In terms of design, a distinction is made between exclusive residential use and the meeting place by the workshop building, where the semi-public character of the complex is palpable. A grove of rock pears and fields of colorful flowering shrubs welcome visitors. Two western courtyards are reserved as private islands of tranquillity for the residents. The development forms niches between the lawns where people can stay and enjoy communal activities.

Depth effect by opening: Garden in Udligenswil

The garden was originally strictly divided into two spaces. The demolition of a dividing wall and new terrain modeling opened up the garden, creating a space with depth. A dry stream meanders from the top side of the property down to a seating area. A massive, shapely wooden staircase leads to the upper garden courtyard, which is not visible from the outside but still offers views of the valley. A gravel area offers space for relaxation and tranquillity, while different varieties of Japanese maples create diversity in leaf shape, foliage, and color. The border planting with bamboo is complemented by flowering rhododendrons and azaleas. Grasses, mosses, ferns and funkia round off the planting, creating an harmonic atmosphere. The yew bonsai in a sandstone bowl provides a worthy focal point in the background of the garden.

Inner Garden

The Inner Garden creates a finely meshed passageway adapts to the new use and adds to the existing range of open space types: A new path is to be laid along the inner plot boundaries through a garden-like atmosphere. As part of the concept, the spatial character was defined and rules for its phased implementation were laid down. Requirements for an ecologically and socially high-quality outdoor space had to be taken into account. The examination of a new open space typology, and the fact that it was not a project but rules for a transformation process that had to be designed, called for new methods. The implementation was challenging from a legal perspective: questions about cost sharing, liability, and maintenance influenced the design process, as did the question of how cross-border implementation could work.

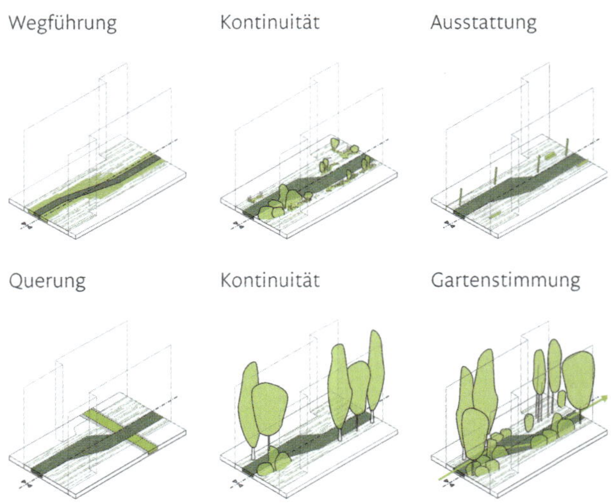

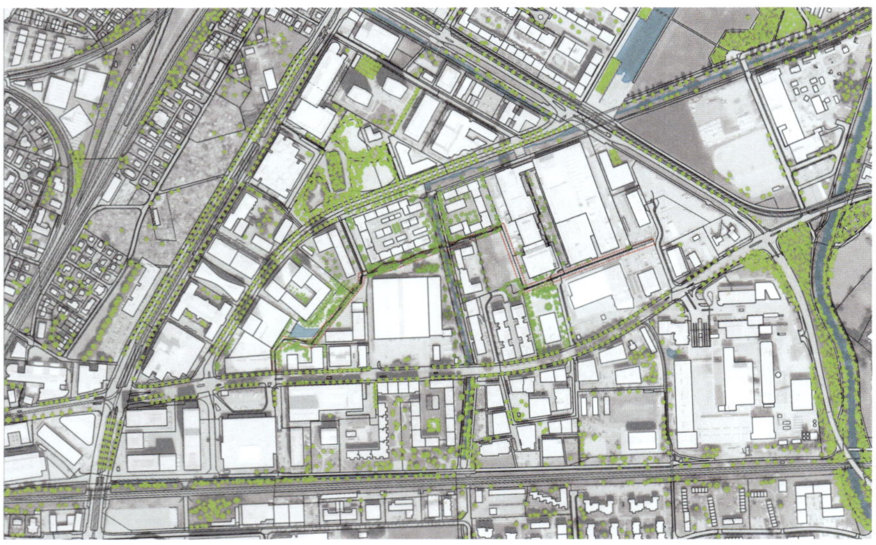

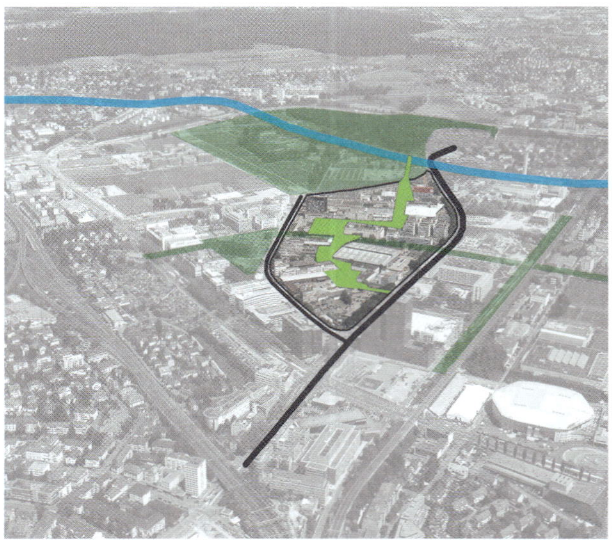

Allmend School Forest Room

On the roof of the Allmend school complex, in addition to the all-weather pitch, there is a school and break area designed as a forest room. With a direct view of the Sihlwald forest, the vegetation is based on it. Playfully curved in-situ concrete seat walls frame the predominantly native forest vegetation. Pines, copper beeches, and hornbeams, as well as a herbaceous layer of woody grasses, ferns, and perennials enrich the school's everyday life with tangible experiences of nature. The forest room is covered by photovoltaic modules like a canopy. Rainwater is also collected here for irrigation. The forest vegetation is occasionally sprayed with a gentle mist of water to create a pleasant microclimate on hot summer days and to enhance the atmospheric forest feeling on the roof of the school building.

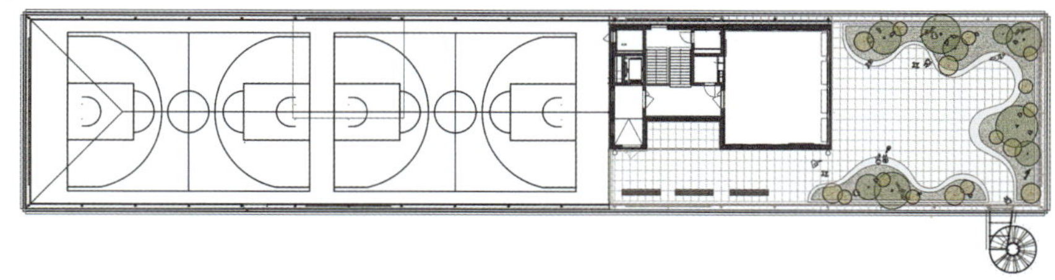

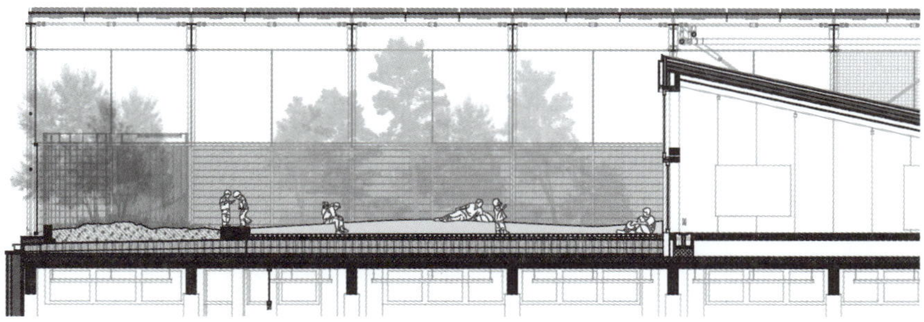

Nature & Business

Since 2011, Die Naturgartenplaner supports the Vorarlberg town of Rankweil in providing near-natural public green spaces. In 2013, an eight-hectare business park was added with the aim of making all company premises nature-oriented. In this project for pharmaceutical company Vetter, lawns and weedy drainage became a species-rich company site with native habitats for a large number of plant and animal species. Attractive spaces were created for employee breaks and festivities. The minimized sealed areas are sown with wildflower lawn on gravel substrate, and rainwater is drained away on site in near-natural troughs. This way, the company premises fight the major crises of extinction of species, loss of biodiversity, and climate change. Nature shows itself in unbelievable abundance in spring-blooming wildflowers.

Gruene Erde Breathing Headquarters

The Gruene Erde Breathing Headquarters' augmented climatic and atmospheric plant performance is applied on a large scale as a bioclimatic ventilation and cooling concept. Planting yards provide fresh air for the entire ventilation of the building. A large meadow in the Almtal valley in front of the headquarters houses a diverse landscape of fruit trees and large native shrubs: 450 trees up to seven meters tall and some 700 shrubs of exclusively native species have been planted in the former corn fields. Two greenhouses grow vegetables for the company's own bistro, and visitors can also see the germination and growth of the plant raw materials used in Grüne Erde products. The surrounding orchards are planted with various heirloom fruit varieties and are left largely in their natural state to allow insects to hibernate.

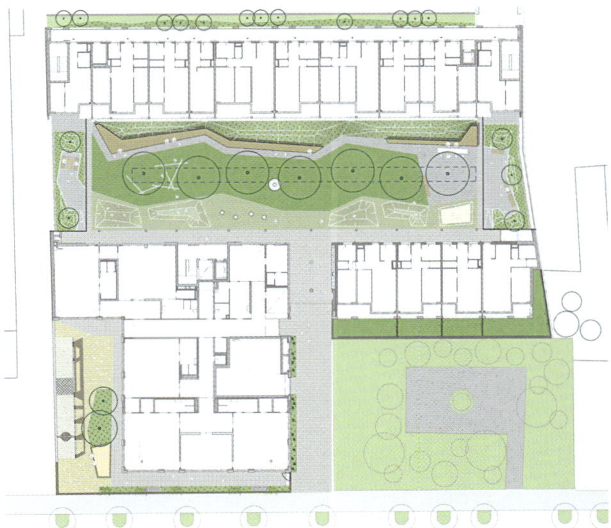

Residential Complex at Brehmstraße

The open spaces of this 127-unit subsidized housing complex include a courtyard and open space for a kindergarten, as well as a spacious entrance zone. Residents also have access to two roof terraces with herb and fruit gardens. The concept of Around the World in 8 Trees has its origins in the multicultural background of the residents and is based on the novel Around the World in 80 Days by Jules Verne. Eight maple trees symbolize the stages of the journey, as maples can be found on each continent. To encourage social contact, the design of the courtyard with its 500-square-meter children's playground was divided into four zones. The highlight of the kindergarten's outdoor area is the playing area between an existing and a new, elaborately designed play wall, which maximizes the play value in a minimal space.

Iles Flottantes

The Iles Flottantes offers the residents of a large complex a shared public space. It serves as an outdoor meeting place, a recreational area, and a playground. Further, it is subsidized by the municipality of Vienna and six different developers and non-profit cooperatives. An upper garden leads from the level of the second floor to a more secluded part of the structure. The first floor is mainly reserved for stores and small offices. The area is defined by floating green islands. The design includes a water area at the meeting place, a boulder wall for children, and a chill-out zone for young people – rare in social housing. A selection of trees, a mixture of flowering shrubs and ornamental grasses, as well as some evergreens create an inviting and aesthetic appearance.

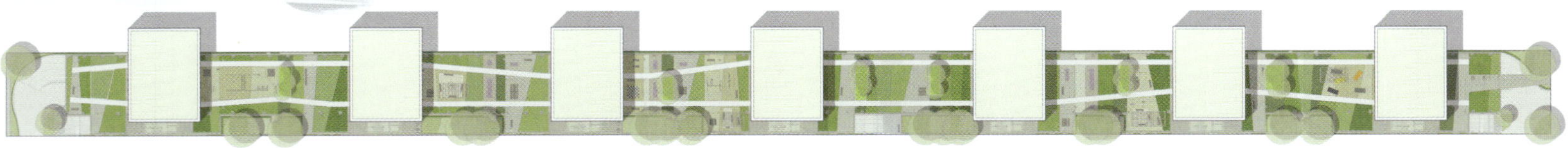

New Community Building: Karlheinz-Hora-Hof

The public housing project is prominently situated on the banks of the Danube in Vienna. It was a particular concern to improve the quality of daily life for the residents living in the neighboring community building. Despite all differences, the existing building from the 1970s and the new building complex form an aesthetic unity that is mutually dependent and complementary in terms of form and function. The seven new building blocks with a total of 332 council flats were smartly positioned and differentiated in heights that continue to allow neighboring residents a good view of the Danube. The blocks are elevated above a 400-meter-long and 20-meter-wide garden deck on the first floor above a parking garage. The garden deck is open to the public and offers differentiated areas with a wide range of activities, framed by a rich green diversity.

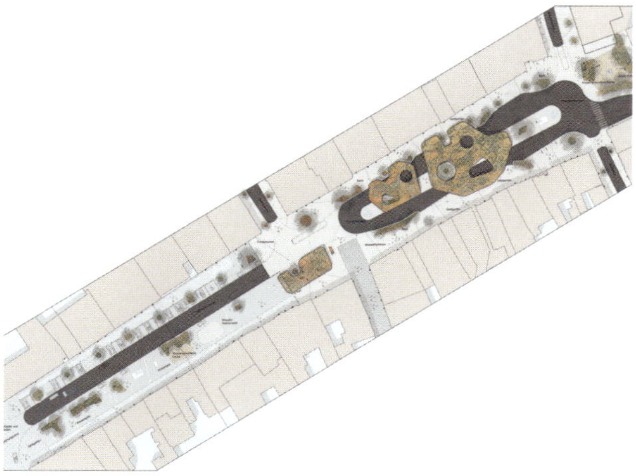

Kaiser-Josef-Platz

Kaiser-Josef-Platz was once a large, 8,000-square-meter suburban square that connected the countryside and the city. Green roofs of varying sizes span the square, protecting it from precipitation and providing shade and cooling. Underneath are the bus stops for eight articulated buses that stop at the same time. The most important feature of Kaiser-Josef-Platz is the new climate-friendly tree and shrub planting. Cooling trees in raised exposed concrete planting islands together with the landscape islands form the framework for the urban space, which also offers space to linger. The lighting creates a bright, friendly ambience in the evening. LED strips under the roof and the bench underlighting of the planted islands lend a special appearance, while two fountains add the element of water to the redesign.

→ U / U studio, Re_place → Education and Playgrounds → 2020 → Prague → Czech Republic → Photos: Jiri Kotal

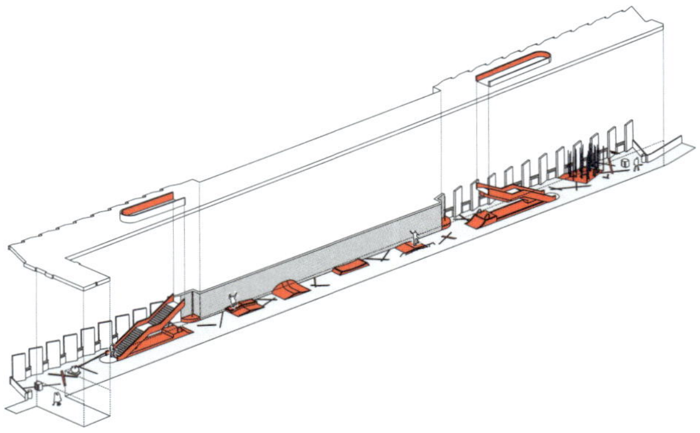

Vltavská Underground

The Vltavská underpass, which had lost its purpose as a transit route between the lanes of Hlavkuv Bridge, was a place avoided by the citizens of Prague. The aim of the project was to replace the vacancy and darkness with a space for freestyle activities in two stages and to create a safe space for its users, pedestrians and cyclists. There is a streetball court, a dance floor with mirrors, obstacles for freestyle sports, a boulder wall, and even a public art gallery. Water elements referencing the linear course of the Vltava River became key aspects of the underground space design. The whole space went through deep cleaning and surface improvement. Steel and concrete for elongated durability were the main materials used. The feeling of safety is enhanced by new linear lights throughout the area.

Green Sgraffito

Within the organism of the historical architecture of the Salm and Schwarzenberg palaces, this place is squeezed behind the walls, an intimate hidden place overlooking a piece of nature in the courtyard. The landscape architects inserted a simple rectangular shape inspired by the morphology of the envelopes on the surrounding façades. The envelope, a black and white sgraffito, is so decorative in itself that they left any other articulation alone, letting the plants stand out in an area of light gravel. From a slightly elevated view of the contemporary gallery space, the effect of the green sgraffito is even more striking. The greenery in the flowerbed is allowed to grow in the form of wild, changing plantings, creating a secret garden for art and tourist-weary visitors.

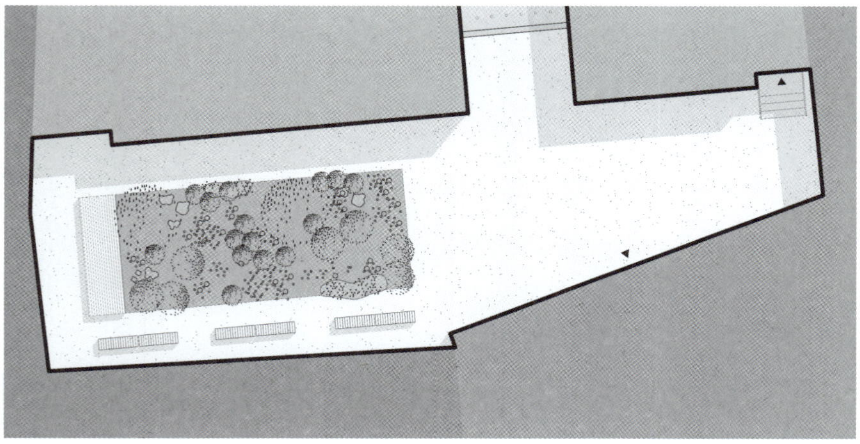

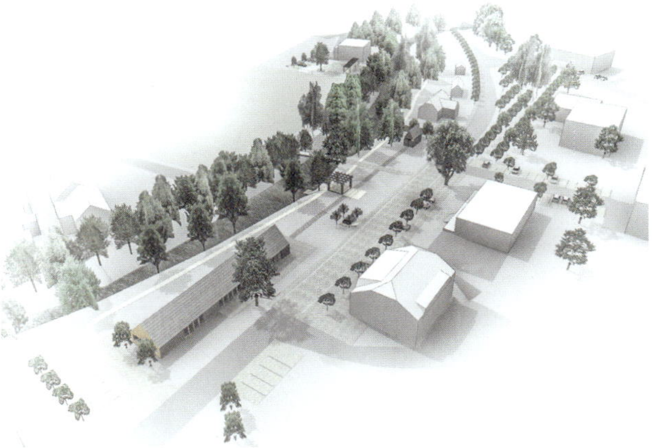

The New Village Square of Prostřední Bečva

The New Village Square serves the municipality of 1,700 inhabitants as a revitalized center, since the village Prostřední Bečva was an overall scattered settlement of shepherds in the 16th and 17th centuries. The solid surface, paved with local sandstone, is complemented by an avenue of trees bordering the street. Between the road and the river is a gentle park landscape with iconic buildings such as the meditation house and the pavilion. The river Bečva is newly integrated into the organism of the square through an open embankment. Materials with a local character were used for the design. Stone from the Beskids, wood, shingles from Wallachian houses, trees from the surrounding pastures, blooming meadow flowers, and rippling water. The square is now also used for traditional events such as fairs, weddings, and Easter.

In Harmony with Nature

The development of the waterfront of Słupca Lake in Poland is a series of circle-shaped bridges, picnic platforms, an integration circle, a stage, and a jetty with a viewing platform. The minimalist platforms on Słupca Lake fit in harmoniously with the natural landscape of the lake and create a space that is friendly to residents and tourists. The project features a simple, streamlined design that prioritizes usability. This is an example of the combination of the design and a respect for the natural environment. The project is focused on smaller objects that not only complete the landscape, but also encourage active use of the space, enabling guests to completely appreciate the beauty of the surrounding natural environment.

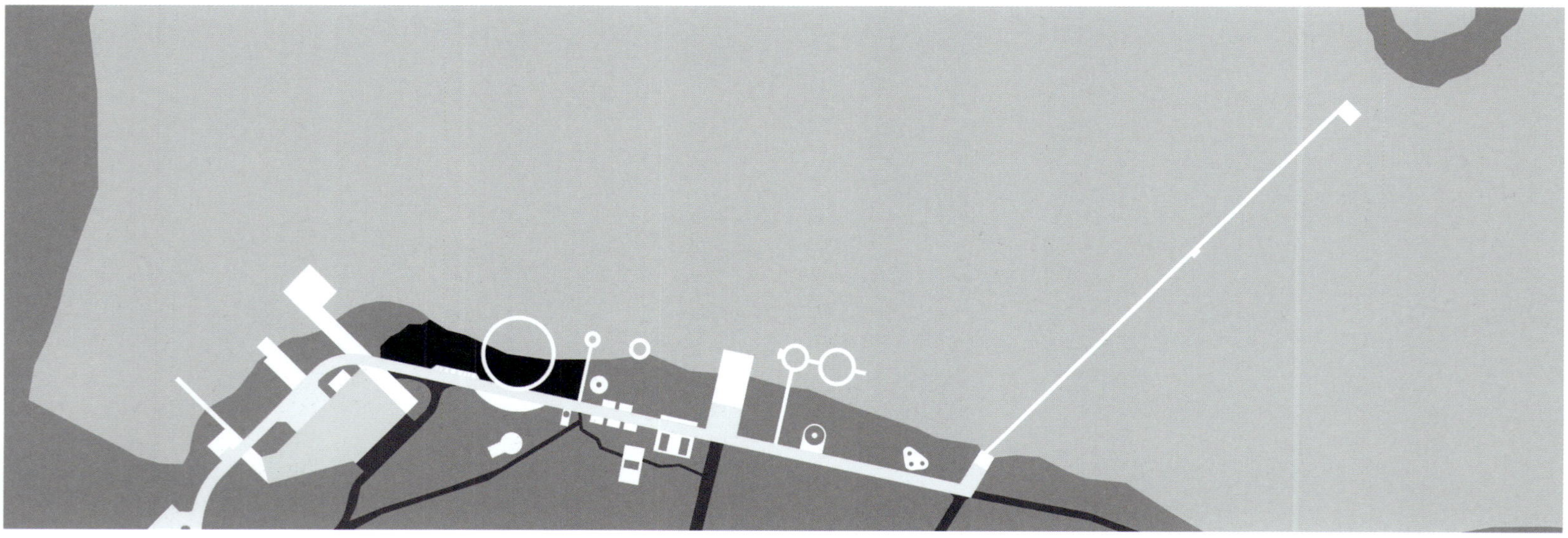

Upgrade of the Pole Mokotowskie Park

Pole Mokotowskie Park is one of the largest and most popular parks in Warsaw, as its space has been shaped equally by the activities of its designers and a grassroots social movement. WXCA architects proposed an evolutionary, not revolutionary approach, a sensitive stimulation of the existing space with its own rhythm of biological and social processes. The biggest change was the naturalization of the water reservoir, which is a popular recreation spot. It was converted into an all-year-round pond where the high quality and purity of the water will be maintained. New urban furniture was designed over the main reservoir in the form of a wooden deck with a terrace layout complemented by wooden bridges and small architecture for various forms of rest and leisure, including seating areas, lounger platforms, and picnic tables.

→ Pupa/Life over space → Public Parks and Gardens → 2022 → Panevėžys → Lithuania → Photos: Aiste Rakauskaite

Senvagė Park in Panevėžys

Senvagė is an old riverbed park in the city center of Panevėžys in Lithuania. After the renovation in 2022 the park was opened to the public. The park with the red brick loop path around the pond has become a popular place for recreational walks. Piers, bridges, and wooden terraces create opportunities to get closer to the water and a little island creates a relaxing environment for small gatherings and sun-bathing. The park is a functional space as well – many paths connect the space with surrounding neighborhoods, upper and lower viewing terraces, a main bicycle route, and the historical city center. With the location and its attractivity for the residents, the park is the heart of Panevėžys.

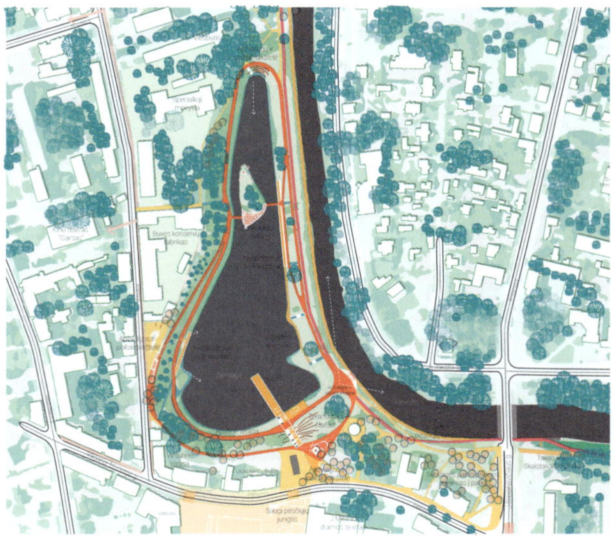

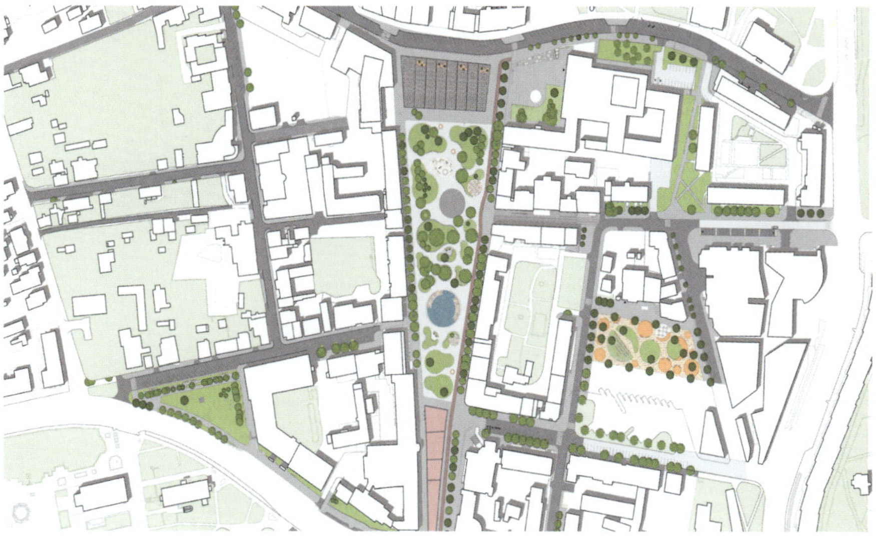

Freedom Square

In 2016, the city council of Panevėžys decided on a major renovation of the main city square. The main idea was to create an open space that encourages the citizens in outdoor activities. Therefore, the design team faced a difficult task – to enhance the pre-existing benefits, such as large open spaces, old trees, and functional pedestrian layout, with subtle architectural innovations, which improve the square's attractiveness to the residents significantly. The historic layout was improved with architectural tools of modern playful design. The architects of Studija 501 decided to divide the central part into smaller islands, enhancing each one of them with specific functions – children playgrounds, planting islands providing calm retreat, and some more private spaces in between those islands.

Ogmios Miestas – Public Space and Urban Transformation

Ogmios Miestas is a gradual transformation of a closed Soviet army base into a thriving, walkable, and fully integrated city quarter. Streets, footpaths, squares, parks, fountains, and a pedestrian-first environment, designed by Do Architects and designer Dalia Stasevičienė, have turned the 120,000-square-meter territory into an attractive place to enjoy leisure time. A variety of functions — shopping, working, catering, and recreation — were chosen to create an atmosphere that is lively, inviting, and diverse. Yet older elements have been preserved, ensuring connection, familiarity, and legacy. The territory has been integrated into the urban fabric of northern Vilnius — both physically and emotionally — and continues to evolve.

→ Artilērijas dārzi, ALPS → Public Parks and Gardens → 2020 → Riga → Latvia → Photos: Aleksandrs Muižnieks, Madara Kuplā, Kristīne Madjare

Gardens of Sporta Pils

Gardens of Sporta pils is a grassroots initiative that has revitalized a privately owned and neglected site and transformed it into an urban community garden from 2020 up to 2023. The aim of the intervention is to create a sense of belonging and a place that serves as a platform for meaningful social change, while promoting a healthier, more sustainable lifestyle based on care and commitment. The site includes 150 gardens at street level and a public park in the pit created by the demolition of building and reclaimed by nature over the years. The groves of the pit also house two contemporary artworks – a painting installation and an artist-run café. The gardens have set a new precedent in the Baltic States for temporary land use and are now completely self-managed by the community.

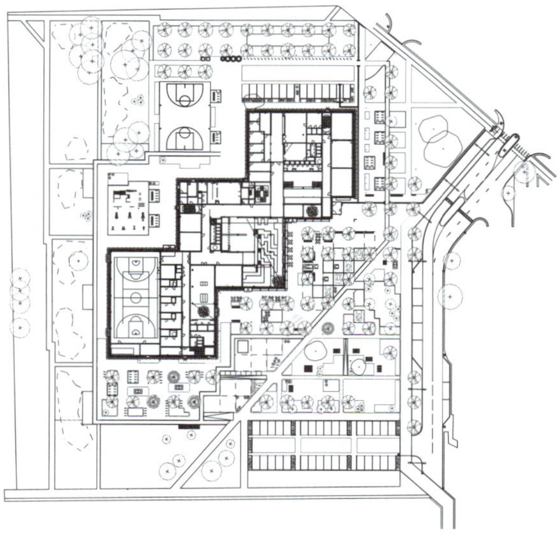

Pelgulinna State Secondary School

Pelgulinna State Secondary School, designed by Arkitekt Must, is a school in a park, but also a parkschool and tree nursery. For the students, who spend most of the day in the school building, the green outdoor area is important, as it frames the interior of the school with calming trunks and treetops. The school's outdoor area is a small living model in which the eco-systems support their inhabitants and vice versa. Biodiversity on the school grounds and in the test areas in the neighborhood is monitored. The plant communities only need conscious care, no constant cutting or fertilizing. All systems are as simple and independent as possible. The basis of the spatial design results from the network from the level of the elements of the building façade. It was the project's aim to preserve the existing and to increase biodiversity by adding new trees.

PLAYHUBarovsk

The city park resembles a city on three hills and with two lowlands: a cozy space for citizen recreation and a huge playground in the middle, the PLAYHUBarovsk. The main element of the design is an extensive, multi-story wall made of red brick. The shape of the wall imitates the natural urban relief of Khabarovsk with many ups and downs and is a direct reference to the most striking historical architecture. As the playground is located near a residential building, it was important to create a quiet zone along the façade. This is why there is a green area here with a narrow, winding walkway that is not accessible to scooters, joggers, crowds of people, etc. This is adjoined by swings and hammocks for children and adults alike. Fruit trees and flower beds lend the area a cozy garden atmosphere, while garlands add a romantic touch.

Bolshevik Factory Courtyard

A heritage-listed former confectionery factory has been converted into offices and high quality apartments. In the courtyard of the residential units, a representative accommodation area was created. Four planted, so-called Dancing Squares invite people to sit on the raised concrete edge, contrasting with planting strips that soften the firefighters' path. A fifth is a playground where children can climb in a ball made of glued laminated wood with a network of ropes. The Dancing Squares motif is continued in the natural stone paving. The project has created a unique green space between strong brick buildings – a modern urban forest that can be seen from all floors. An oasis of calm for residents and employees in the middle of a busy city.

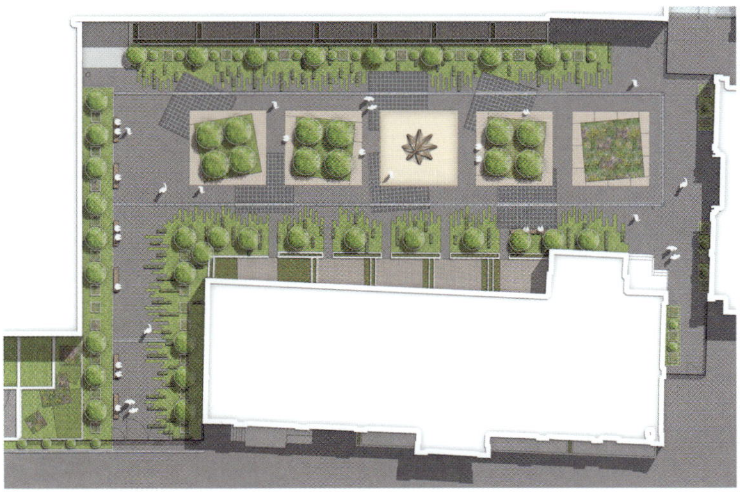

Pedestrian Boulevard Pivdennuy in Dnipro

The boulevard, called Yujnui, is located in the city center and is divided into two zones: Transit areas along the buildings and a multifunctional zone in the middle with recreation areas, rest areas, and a children's playground. The main dominant landscape feature of the new pedestrian area is the stream that runs across the boulevard. It symbolizes the hidden underground river that flows directly under the street. As it is not possible to uncover it again, it was decided to install a river as a foot mark, built from massive blocks of local granite. The length of the stream is 200 meters, making it the longest artificial stream in Ukraine. A seven-meter-high bronze Atlas sculpture by Egor Zigura rises above the stream. In Greek mythology, Atlas carries the firmament on his shoulders and symbolizes liberation from limitations.

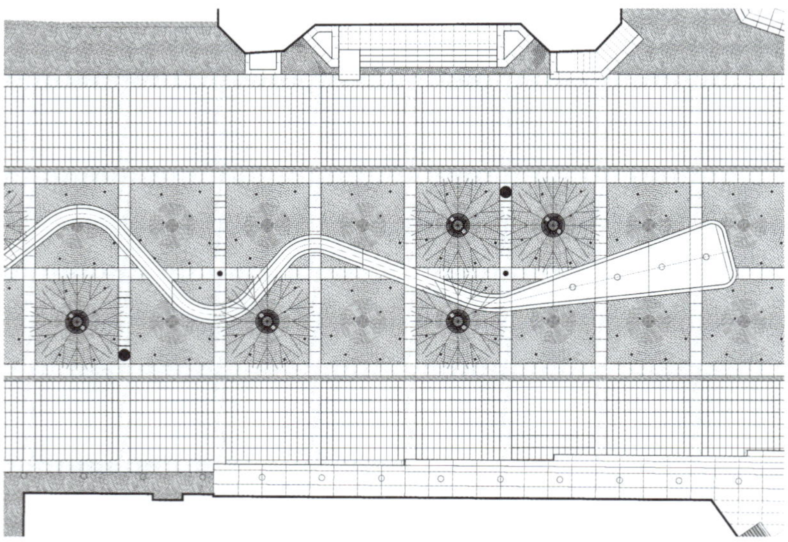

Inventor Playground

School is the place where children spend a lot of time, which is why a creative atmosphere is important and can enhance the desire for ambitious achievements. The project aims to realize creativity, brightness, and consideration for the comfort of children in this colorful school playground. The design was developed in a non-standardized way regarding the concept to make the school standout and to be remembered for its unique identity. The basis for the development of the landscape was a famous mural called "The Mural of Unusual Size" by Alex Brewer (known as Hense). He transformed an abandoned industrial building in Maryland into a piece of art with his large, colorful mural that is now the most popular spot in the state. The Inventor Playground is now a popular space for children to socialize, interact and play.

River Mall Embankment

This project entails the renovation of a section of the embankment, located on the left bank of the Dnipro River in Kyiv, Ukraine, which became a significant part of the city. The renovated embankment spans 300 meters, offering a circular layout that encourages long strolls and diverse activities. LDesign aimed to create a modern and inviting space that caters to social needs for relaxation, socializing, and active leisure. Their objective was to design an area where people would willingly spend their time. This green oasis was anticipated to become a popular community hub. The architects as project creators envisioned a captivating destination where people can relax, contemplate, and enjoy social gatherings. Judging by the considerable number of visitors, even during colder seasons, it seems the layout achieved this goal.

Koliivshchyny Square

Koliivshchyny Square is the smallest square in the center of Lviv with its own unique character and atmosphere. Historically it was a place of exchange in the old Jewish quarter, today it is one of the most visited locations in the city center. The most important tasks during the redesign were to preserve the scale of the square and its unique atmosphere, to identify multiple layers of history, and to visualize them. The 19th century well, remains of buildings that no longer exist, and the remains of the defensive wall of the old city were reconstructed on the basis of archaeological excavations. In memory of the old residents, their names and professions are carved on the edge of the restored well. After the reconstruction, the square became more accessible and inviting, but kept the old layout.

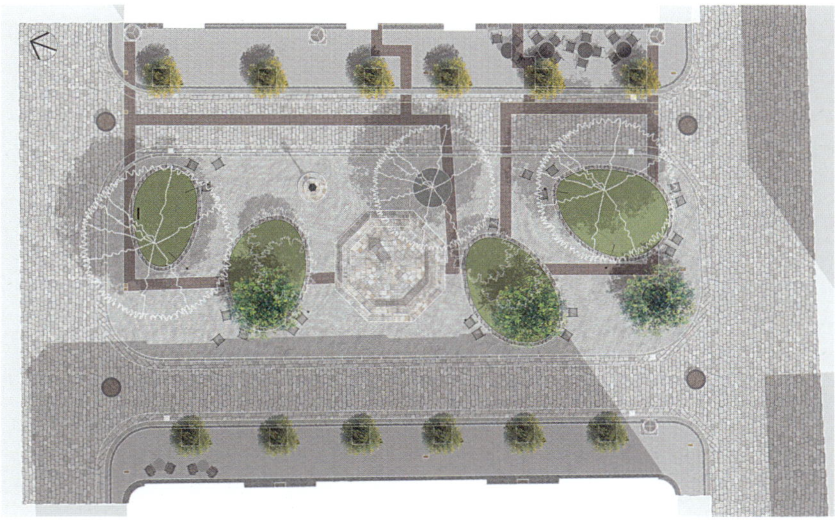

Garden of Communities in Veszprém

Placed in the centre of the Factory'ard in Veszprém, the Garden of Communities, designed by Hello Wood, is an artwork that celebrates the rich culture of the Veszprém-Balaton region. The pavilion is a symbolic monument to the cooperation of the 116 municipalities participating in the Veszprém-Balaton 2023 European Capital of Culture program. Each municipality has a unique planter box, with the name displayed on the side, containing a sapling. The uniquely shaped planters create a spectacular exhibition space, accessible to anyone. The contrast between the geometric architectural forms and the lush vegetation creates a magical atmosphere, attracting visitors who can wander around the pavilion and browse through the names of the villages or cities that hold dear memories for them.

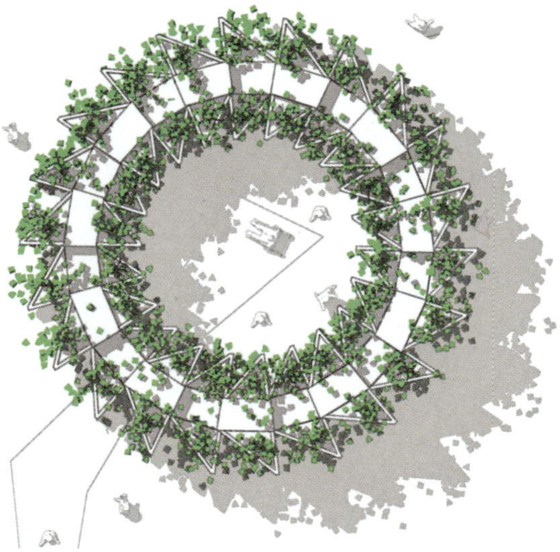

ALSÓÖRS
AJKA
VESZPRÉM

VINDORNYAFOK
ZALAHALÁP
BALATONUDVARI
NAGYESZTERGÁR
VÁROSLŐD
TAPOLCA
VÁSZOLY
ALSÓPÁHOK
BALATONBERÉNY
BADACSONYTOMAJ
BAKONYNÁNA

The Promises of Modernist Premises: Pool Trakiya

This collective re-appropriation of a deserted pool as a communal outdoor center reclaims space between block 1, 2 and 3 of the large prefab estate Trakiya in Plovdiv, built in the 1970/80s. The pool – dry for almost the entire time of its existence – only remained in physical traces and could not be reinstalled. In this kind of modernist housing estates, there is vast in-between space which belongs to no one in particular and to everybody in common; it is a space beyond ownership and individual home. The transformation collectively reactivated the place from its fringes. It was carried out with the help of the residents, became a valued area for exercise, dance, debate, play, and meeting, and became the driver for re-installing public water amenities. The redesign also triggered collective care-work for the shared public space.

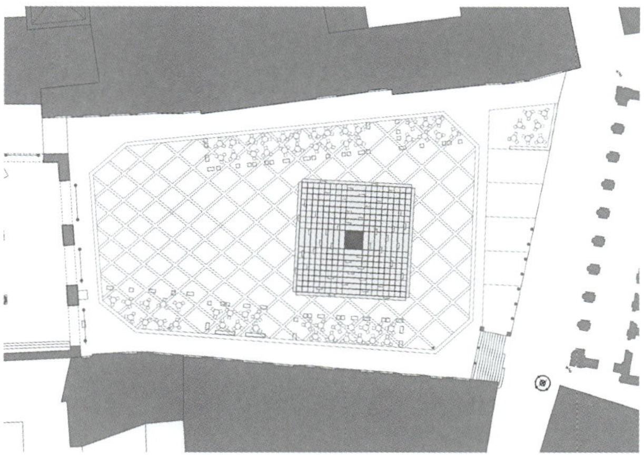

Grow Together Grow Green

Situated in the main square of the old city of Bergamo, this installation in the context of the Landscape Festival 2023 wants to show that global challenges, such as climate change and deforestation, can only be tackled through collective efforts. The installation thus provides a physical and figurative framework for joint efforts to improve the urban climate, in line with the Festival 2023 motto: Grow Together. Visitors can register to take the seedlings of local plant species, creating a sense of engagement and responsibility. The modular structure embodies its own philosophy, being made of scaffolding that is light and easy to assemble and disassemble, and therefore has an indefinite life cycle. The work also connects with the surrounding urban environment, offering visitors a view of the historic old town.

Casa Olearia Paigle

In the serpentines high above Gardone Riviera, two buildings connected by a glass structure, reminiscent of the ubiquitous limonaia frameworks, are nestled into an olive grove. The water basin blends in with the colors of nearby Lake Garda. The 324-square-meter private vacation home is framed by dozens of newly planted olive trees, while white oleander and rosemary accompany visitors visually and in an olfactory way on their way to Casa Olearia Paigle, a minimalist building with typical local skeleton construction, using natural materials such as light maple wood, Thassos marble, glass, and plaster, as well as individually designed and manufactured furnishings, accompanying the landscape that changes with the light.

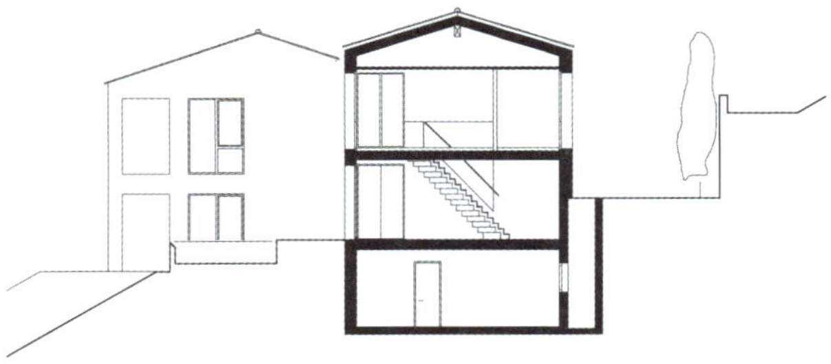

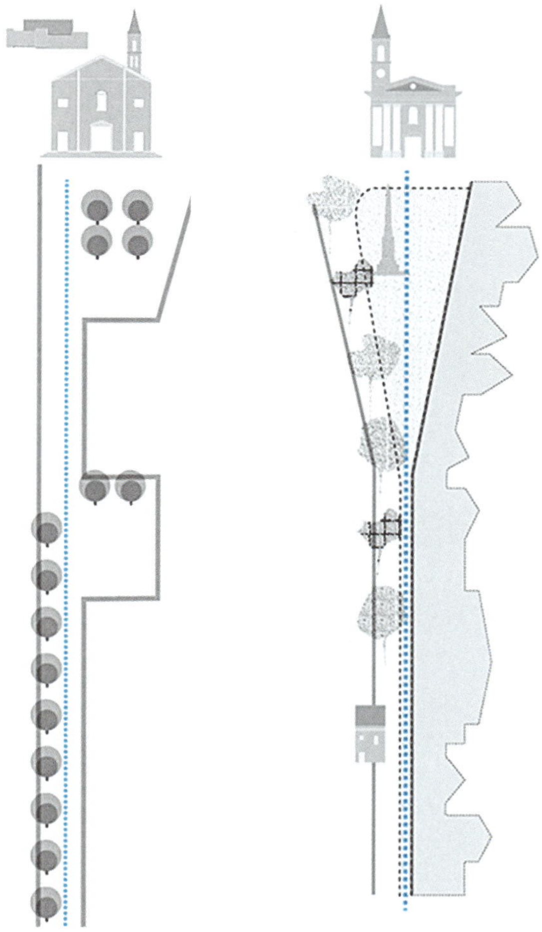

Urban regeneration: Piazza San Rocco and Corte Sant'Ilario

The project is located in the ancient village of Borgo San Rocco, near Gorizia. The purpose of the intervention was to revitalize the historic central square of Sant'Antonio. By emphasizing the perspective axis of the church as a focal point it connects the square with the churchyard through a selection of ideal materials. Existing artifacts, various spaces, points of lights, as well as memorial plaques commemorating important individuals, were integrated into a unified ground design. The square extends to the built curtain and connects the open and private spaces using continuous paving that incorporates color and texture from the surroundings. The project expanded to the surrounding streets including the renewal of the road surfaces, the use of porphyry paving for the sidewalks and the creation of a new vegetation structure.

Czech Pavilion Rooftop

Steppe and rock formations mark the beginning of the exhibition in the left portion of the entry area. The steppe plants represent the high-tech system of the third millennium, transitioning from their native plant habitats into beds of crushed glass mulch. Simultaneously they are a reminder of the rich history of Czech natural medicine, a once traditional but now alternative way of treating diseases. Glass is a primary component of the garden and symbolizes the legacy of Czech glassmaking, while also having a wide range of future uses. A raised planter with trees and shrubs separates the prairie nature from the following meteorite area. The scenery shows the impact of a meteorite that has created a barren landscape, where plants struggle to survive in the inhospitable environment, recovering further away from it.

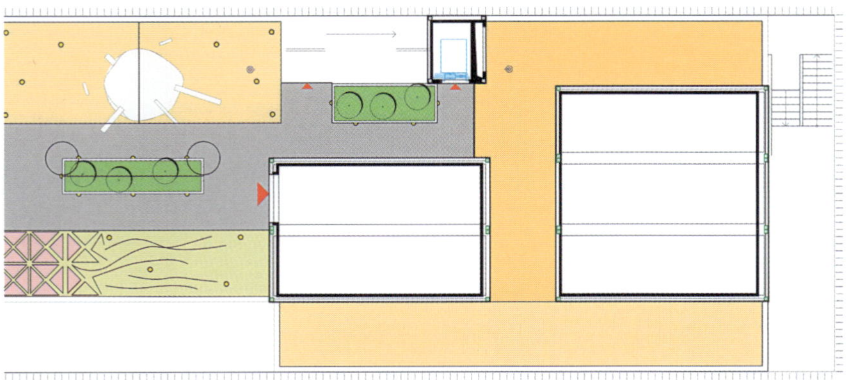

Nutura – Re-experiencing Nature

This pavilion, surrounding an imposing Gingko Biloba tree, was built for Roca as part of the FuoriSalone in the courtyard of the Università degli Studi di Milano. The name of the pavilion is a blend of Roca's latest collections, Nu by Inma Bermúdez and Tura by Andreu Carulla. The installation reflects the lifestyle of the company, which originates from the Mediterranean, where light casts powerful shadows and dazzles in its land and sea. A platform in the shape of the given plot raises the installation above the surrounding hedge. The different heights and levels of transparency of the undulating walls permit a glimpse of the internal paths and allow the visitors to approach the tree. The special ceramic tiles that cover the floor are a playful way of presenting the new Roca collections.

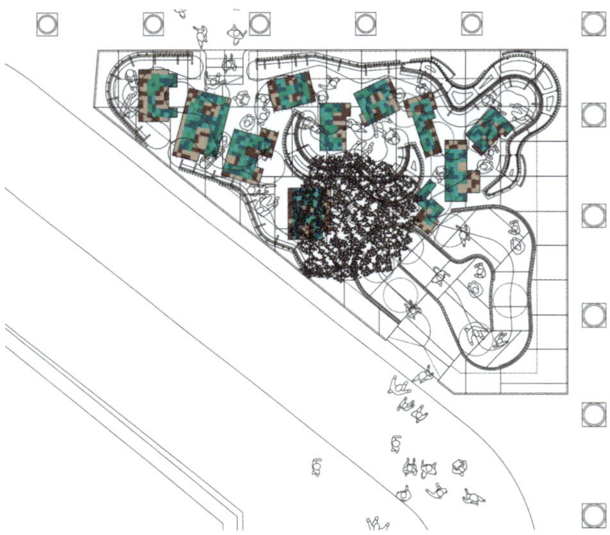

New Square in Via Campagnola

The project redeveloped a redundant building in Via Campagnola and the adjoining small religious chapel at the Via San Marco. The main aim of the project was to increase pedestrian and traffic safety, and to redesign the area. The abandoned building on the main road was separated from the chapel and demolished, while the chapel itself was carefully lifted and moved to its new location. On the site were the building once stood, a new square was created with a green area on which the chapel now stands. In this new public space, worshippers can safely perform religious acts and people can rest on the prefabricated concrete benches in the shade of the Ginko Bilobas surrounded by planting. A new brushed concrete sidewalk improves usability and safety for pedestrians, connecting the new public space with the adjacent existing church.

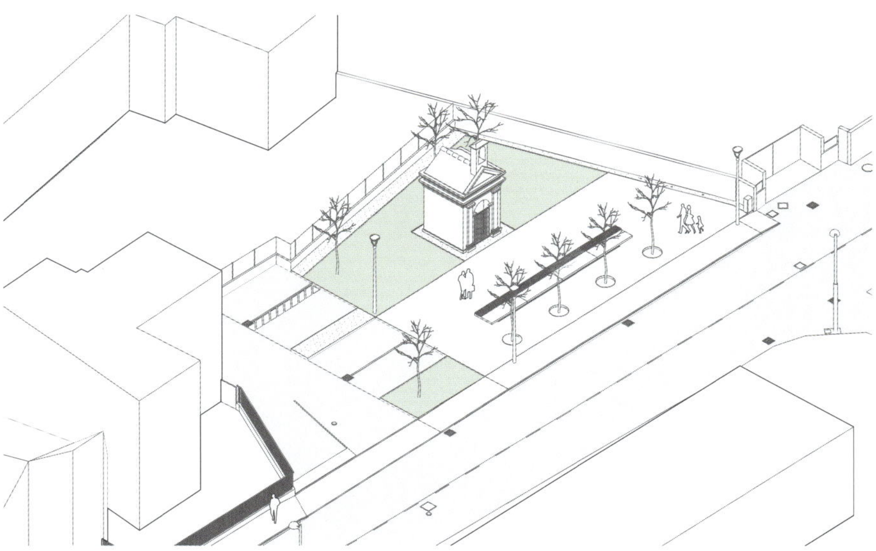

Between Woods and Vineyards: Boscareto Resort & Spa

Nestled among the vineyards of the Langhe hills in Piedmont, Boscareto Resort & Spa has re-opened its doors to welcome guests in a new green setting. The landscape design was inspired by the surrounding vineyards and flora, while taking into account the needs of the resort. As the Langhe region was declared a Unesco World Heritage Site in 2014, the resort is often visited by experts in the field. To cater to this group of guests, the large terrace, which offers a breathtaking view of the Langhe hills, has been redesigned. Vegetable pockets were created, polygonal shapes filled with greenery that visually extend the surrounding vineyard fields. Here, guests are offered a unique setting to relax while enjoying the surrounding scenery. The greening was designed by Hw Style, the greenwall by Harpo.

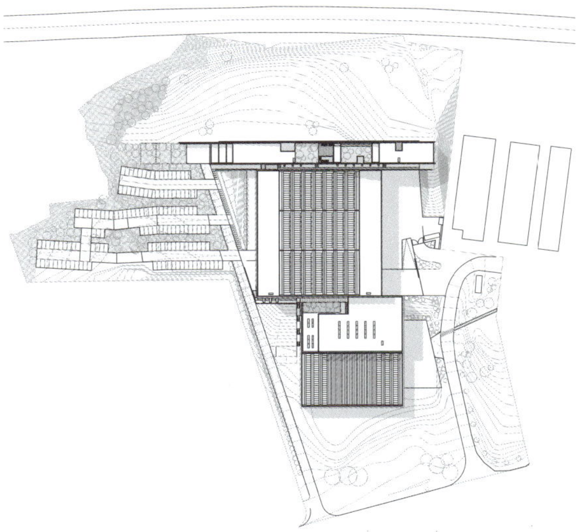

Furla Headquarters Progetto Italia

A key feature of this industrial architecture and office project was its integration into nature, where working environment, sustainability, and landscape are harmoniously conceived. GEZA Architettura was responsible for Architecture Design and Creative Direction, LAND Italia for landscape design. Located in the Chianti area, the Furla factory has three main volumes and three fundamental elements – the driveway, the terraces and the Piazza Furla – which define clear areas around the building volumes. A system of patios, porches, and green roofs blurs the boundaries between inside and outside. The main entrance recalls the archetype of the Tuscan villa entrance: A linear row of cypress trees greets visitors and points the way through. This rhythm is repeated on the façade of the buildings with solar filtering louvers.

Koper Central Park

The Central Park of Koper is located on the edge of the historic city center and meets the needs of the residents for green spaces and a connection to the sea. The basic building blocks of the new city park are monolithic, undulating urban elements. Their controlled placement in the space, combined with a pronounced topography of green areas, gently divides the entire park area into individual introverted program islands. The design of the individual urban elements consistently follows the different programs. In some places they are higher and serve as a shelter from the view and the noisy surroundings, in other places they are lower and allow an unhindered contact between the program zones. The planted part of the landscape design is made up only of indigenous Mediterranean vegetation.

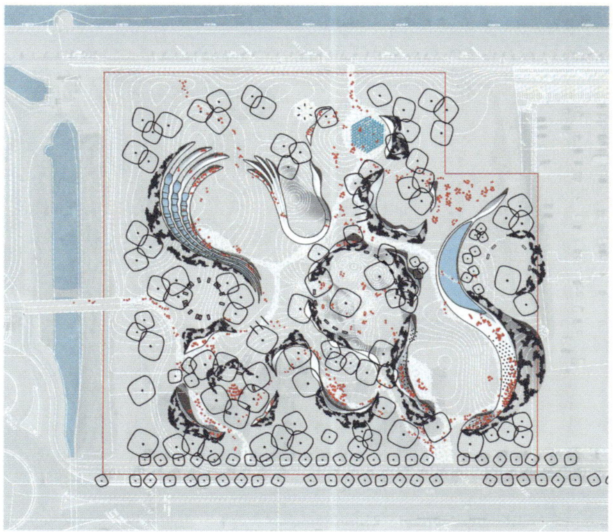

Central Squares in Koprivnica

By accepting bicycle traffic as an inevitability in the functioning of a new city square, bike lanes that form an X establish a sort of coordinate system for the new public space. A tapestry of smaller zones structured around the axes of the bike lanes has been laid on ground. Each zone corresponds directly with the surrounding context or brings its own new amenity: In front of the gallery, there is an art zone with pedestals for occasional exhibitions and circles for drawing on the floor. Under the park trees is a pop-up zone for summer events. In front of the bars and restaurants are café zones for parasols and tables. By using different urban equipment, different greenery, and different paving, each zone has it's own identity, thereby adjusting to the ever-changing needs of the citizens of Koprivnica.

Grand Park Hotel Rovinj

Grand Park Hotel Rovinj is a viewpoint for the most beautiful stage on the Adriatic. In order to blend into the existing forest, the floors offset to follow the slope of the site and form large terraces, Mediterranean gardens, pools, and sundecks. A bar with a terrace and an á la carte restaurant are located in the lobby and the reception area, which is connected to the main restaurant with a central staircase. Great significance in the hotel's view was given to five pine trees that have been growing on the site for over half a century. The areas alongside the promenade are functionally and formally different from other spaces. A relaxing atmosphere is created by water surfaces, carefully selected paving, and greenery, which frame particular content zones and make the whole area both attractive and accessible.

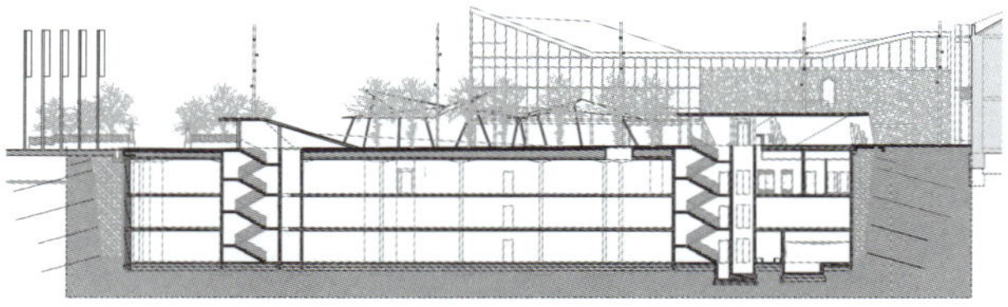

Poljana Square

Poljana Square is designed as a diagram between past and future, abstraction and reality, life and art – a continuous three-dimensional urban platform, which accommodates various social events and mediates between historical and modern urban artifacts. The proposed strategy is open-typed, establishing a dialog in the indefinite field of the modern city, preserving the existing identity, and directing the future use towards new scenarios in a discrete transformation. The surface of the square is superimposed on two networks, a classic grid and a flexible square field of points. In network intersecting areas, spatial folds form extended, raised surfaces on the roofs of the pavilions. The canopy provides a connection and defines the spatial boundaries of the square, creating a dialog and continuity of forms in the new age.

Eco Park Durres

Eco Park Durres is a large public park built on a former contaminated landfill. The design is inspired by a famous local Roman mosaic, called "The Beauty of Durres", which serves as a metaphor for the transformation of a polluted site with negative connotations into a public space for recreation and environmental education. The hills are made of compacted waste. They are covered with vegetation and sculpted to create a land art intervention. A network of winding paths makes the whole landscape accessible, transforming it into a public park for the city, equipped with recreational and sporting facilities. The largest hill provides the highest climbing wall in the Balkans. The Eco Pavilion at the entrance houses a permanent exhibition that promotes ecological awareness among visitors.

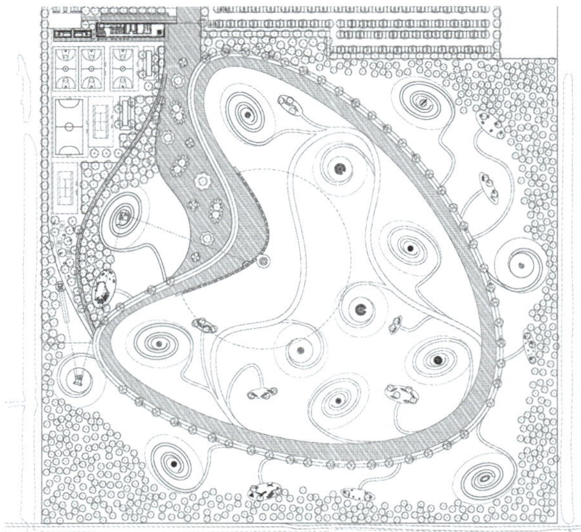

Olympic Course at Costa Navarino

Costa Navarino, in southwest Peloponnese, has grown into one of Europe's leading golf destinations and is regarded as one of the prime sustainable touristic developments in the Mediterranean. The offer includes four 5-star deluxe hotels and four signature golf courses, among them two new 18-hole golf courses, the International Olympic Academy Course and the Hills Course, which opened in 2022 and were both designed by two-time Masters champion and Ryder Cup legend José María Olazábal. The two new courses offer breathtaking views across the Bay of Navarino and stretch over a stunning 1.40 square kilometers of garrique landscape in rolling hills along a cliff-top overlooking the sea. Playing to Par 72 and measuring 6,351 meters, the Olympic Course offers championship golf amidst mesmerizing panoramic views.

Bayou Villas

Bayou Villas is a part of Antalya Lara Barut Collection Hotel. It consists of an entrance building, a restaurant and 25 villas of five different types. The landscaping with more than 200 species of plants, including species such as Strelitzia regina, Strelitzia nicolai, Cycas revoluta, and Alocasia, transforms Bayou Villas into a Mediterranean-like oasis. Each villa consists of interconnected exotic gardens and water gardens. The layers of water and green create a strong language and a sense of character and belonging. The use of both salt and fresh water provides different bathing environments and contributes to the creation of an integrated natural landscape and ecosystem. The entrance building is interpreted as a gateway that separates the new complex from the hotel and opens to a new experience.

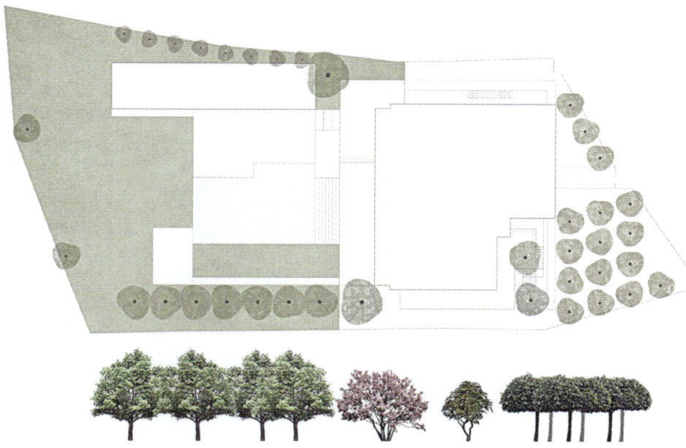

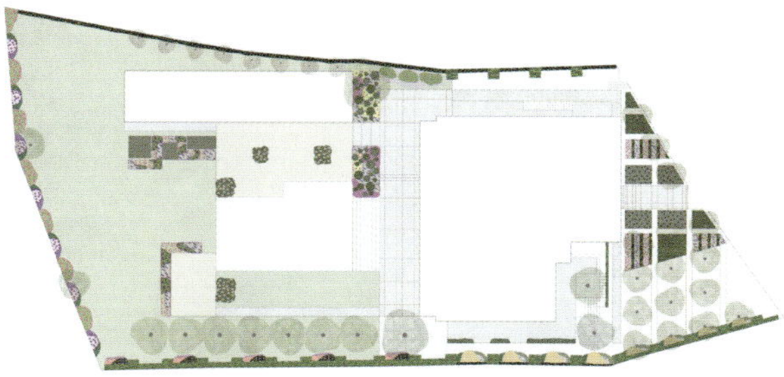

Bahçeköy Villa Istanbul

The garden design is assembled as a series of frames, a composition of thematic elements that relate to the dimension and form of the surrounding environment. Each setting maintains the continuity of space within each area of the garden, creating an overall framework that balances representational space and functional space in relation to the built architecture. The thoughtful arrangement not only enhances the visual appeal of the garden belonging to this Istanbul villa, but also ensures a harmonious connection between its various elements and the built environment. Through careful design, the garden becomes a seamless extension of the architectural landscape in which it lies, providing both visual pleasure and functional utility.

Great Mosque of Algiers

The new mosque on a 20,000-square-meter hillside above the bay of Algiers hosts 120,000 visitors daily. The open spaces fulfill several functions: As parts of the religious complex, they create a network of paths between the different areas, provide attractive places to linger, and offer shade and tranquillity. The adjacent landscaped park is formed by paths in the style of Islamic calligraphy. On the west side there are architecturally shaped new Islamic gardens, characterized by hedges based on North African Islamic ornaments. By merging open spaces into a strict grid of paths and squares, the design meets the challenge of respecting and merging religious, cultural and social uses. These open spaces, in which completely different interests and requirements are contained and combined, allow visitors to identify with this sacred site.

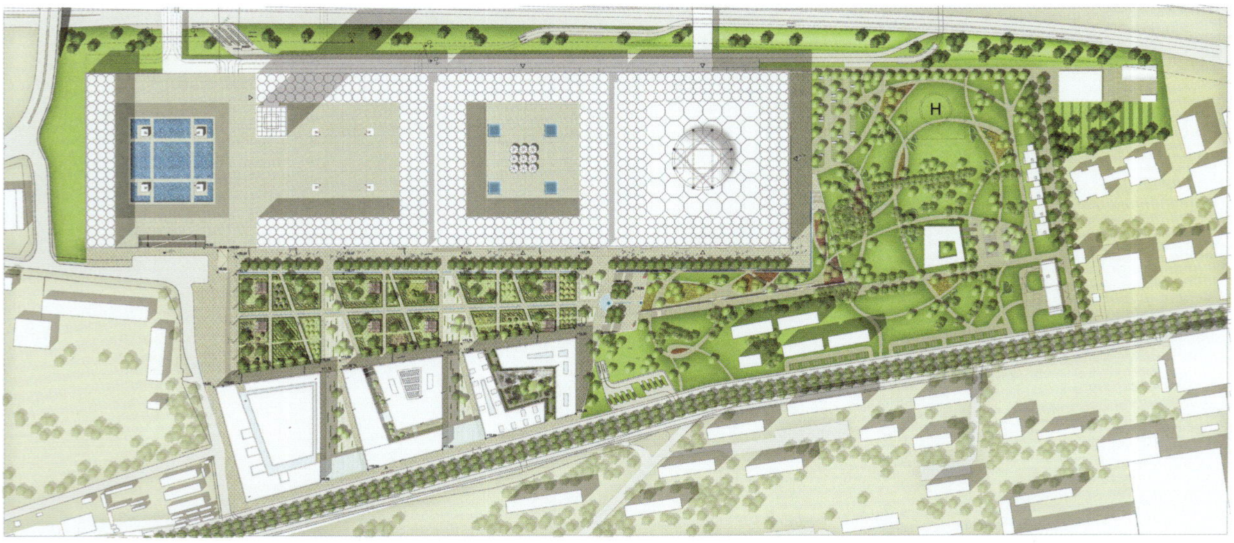

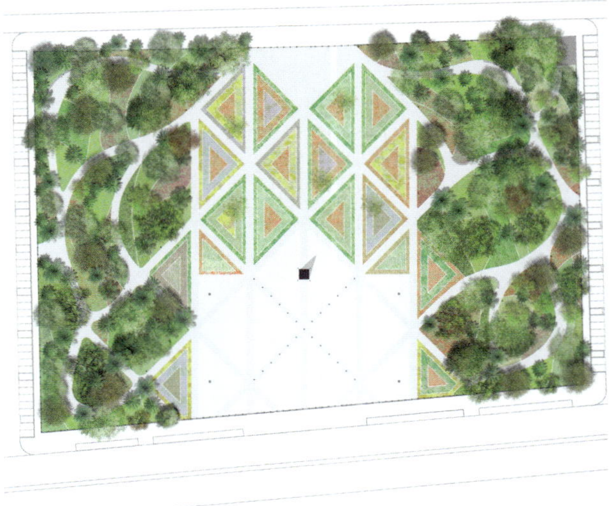

Jardins de Mathieu

The Ministère du Cadre de vie et du Développement durable of Benin had the park built on the former residence of President Mathieu Kérékou to create additional green spaces, reduce paving, and implement greening measures to prevent heat stress. The 30,000-square-meter layout was inspired by beautiful African textile patterns. In the middle of the park is a formal area designated for ceremonies and wreath-laying. It is accompanied by centrally located green flower beds that transition on its sides into a city park with raised, meandering walkways and shaded seating areas nestled beneath the trees. The park features a wide variety of local, dense vegetation that promotes biodiversity. A central obelisk-shaped monument pays tribute to those who lost their lives during the colonial period.

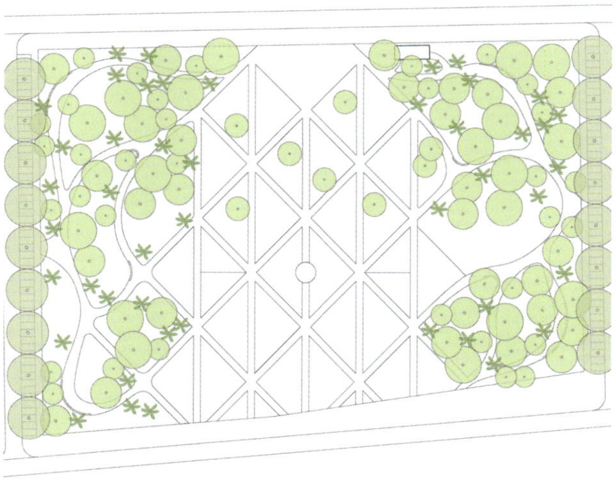

Bosjes

Bosjes is a wine farm in the foothills of the mountains near Worcester. The aim of the project was to make the farm a drawcard for visitors, an oasis with water as a strong theme. The gardens anchor the iconic chapel, subtly depicting events from the Bible. All plants, materials and labour had to be locally sourced. Through the lush garden, the chapel connects to an existing historic graveyard and, through vineyards, to the original Bosjes Manor House. Endemic and biblical plants, such as olives, almonds, cypresses, honey locusts, citrus, and pomegranates were propagated for the project. Materials used are traditional to farms in the cape and include white-washed walls combined with slate and packed stone, reflecting the colors of the mountains. Bosjes is accessible with signage in braille and wheelchair accessible pathways.

A Magical Setting

With the natural setting being the priority, the designers had to decide where to let nature take over and where to tame it. The pathways, walkways, and promenades hint at undiscovered little islands of pleasure. Even the pool house is hidden, allowing for an uncluttered field of vision, where nature and everything it has to offer can truly be enjoyed. The primary axis starts at the house and bisects a natural amphitheater formed by protected rock formations, inside of which lies the infinity pool, gently guiding the line of vision towards the horizon. Walking around the pool, one is taken away into mysterious walkways that tempt curiosity and guide the wanderer into groves of oak, cedar, and fruit bearing trees, making for a truly multisensorial, multi-seasonal experience.

Common Ground – A collective garden

Common Ground is a collective effort to transform the exterior of the historic Beit Harishonim House into a cultivated garden. The construction process for Common Ground took place in an open-to-all construction site. Using open-calls and publications, nearby residents have joined the building process; together with the designers they tightened the ground, casted concrete benches, and planted wild plants. The layout of the garden is a condensed, pocket-sized English garden, offering views of the surrounding landscape, while making room for a bold, custom cast concrete terrazzo element. Nearly all plants are local wildflowers, while cultivated plants have been used to provide multi-seasonal blooms. Once completed, the garden will be taken care of by local residents, turning it to an ongoing project and collective effort.

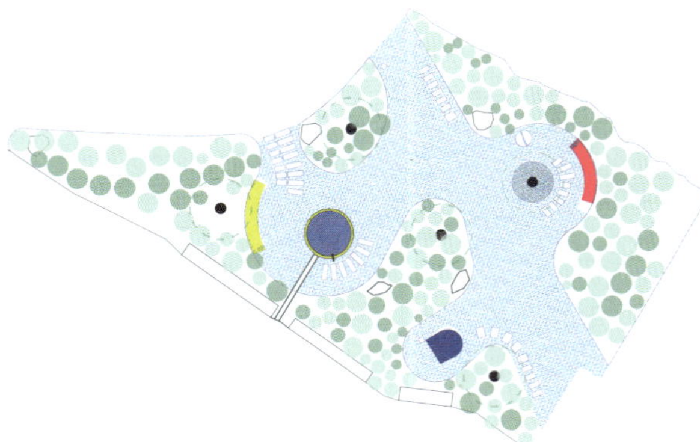

Tavor Park

The goal was to unite the gardens of Kiryat Yam's Dalet neighborhood that extended over 9,000 square meters to create a continuous space that would function as a place of encounter, play, and rest. The route of the 240-meter concrete and asphalt path with stone walls and platforms covered in wood was laid down, linking all the yards without disturbing the system and connecting the exits from the housing blocks into the new garden, along with seating areas, creating open squares and a transitional area. Mature trees were conserved while dozens of new shade trees were added. The planting program comprised useful, water-saving, and local plants, in addition to planting grassy hills. The landscape design with a contemporary layer of installations transformed the run-down space of housing projects into an inviting focal point.

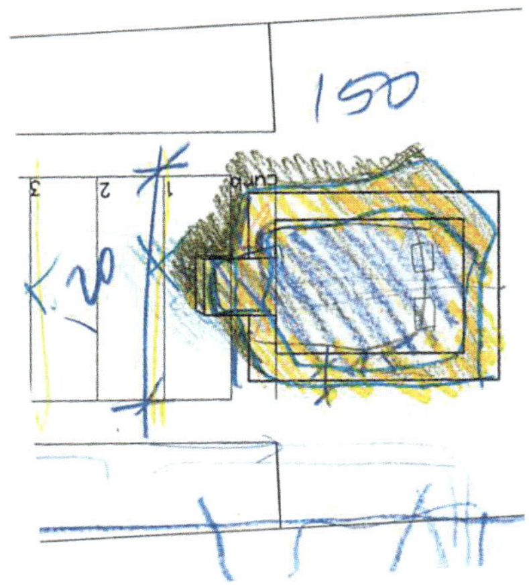

The New City Center of Modiin

The concept for Modiin's new city center, based on Moshe Safdie Architects' statutory plan, envisioned the creation of a vibrant mixed-use central business district, inviting people to discover urban living, working, and shopping along the urban boulevard and green spine of the city's new civic center. The landscape design puts pedestrians and bicyclists first, providing generous space to shop, stroll, play, or rest. Four rows of Plane and Jacaranda trees form the green structure of the boulevard ending in the Light Sculpture Plaza. The central linear promenade includes water features, shade structures, sitting areas, five restaurants, and coffee shops. The sidewalks and promenade are aligned by continuous planting beds which act as rain gardens in the winter months, while providing optimal growing conditions for the trees.

Al Fay Park

Al Fay Park in Abu Dhabi represents a paradigm shift in how to design and implement city nature in the dense megacities of the Middle East. The 27,500-square-meter park is the first urban park in UAE focusing on biodiversity, social activities, and microclimate optimization. Aesthetically and climatically it is designed to be a comfortable social space, utilizing the latest research in microclimate optimization and urban sustainability. Its high density of trees, and modern irrigation systems, thus ensures that it uses 40 percent less water than traditional parks while providing a cool, comfortable, and unique around-the-clock place for all residents and visitors to meet, play, and live. As the city's first biodiversity park, Al Fay Park contains more than 2,000 native trees and bushes, strengthening biodiversity for both plants and wildlife.

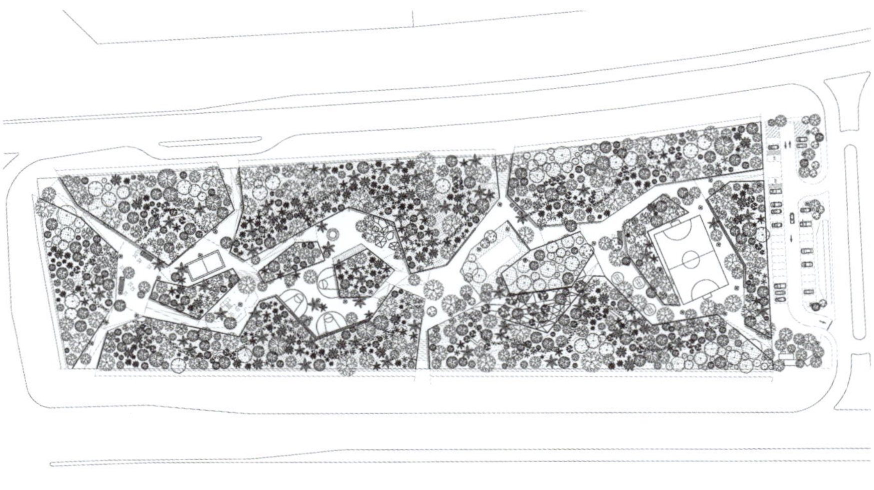

Oxygen Park, Education City

Inspired by the elemental unit of life, oxygen, Oxygen Park is a public space created in a desert climate designed by Erik Behrens and James Haig Streeter with the AECOM Landscape Team. It is designed to serve the community and its well-being by providing opportunities for rest, exercise, and play. It is inspired by the wind-eroded rock and dune formations of the Qatari desertscape. The path and running tracks blend seamlessly into the topography of the park, creating a training ground with looping paths, cooled tunnel segments, and steep hills. A special night lighting scheme and refreshing water features provide an attractive setting for evening exercise in the park. The park also features team sports courts, equestrian facilities, and recreation areas with a series of refreshing, sound-filled folly spheres.

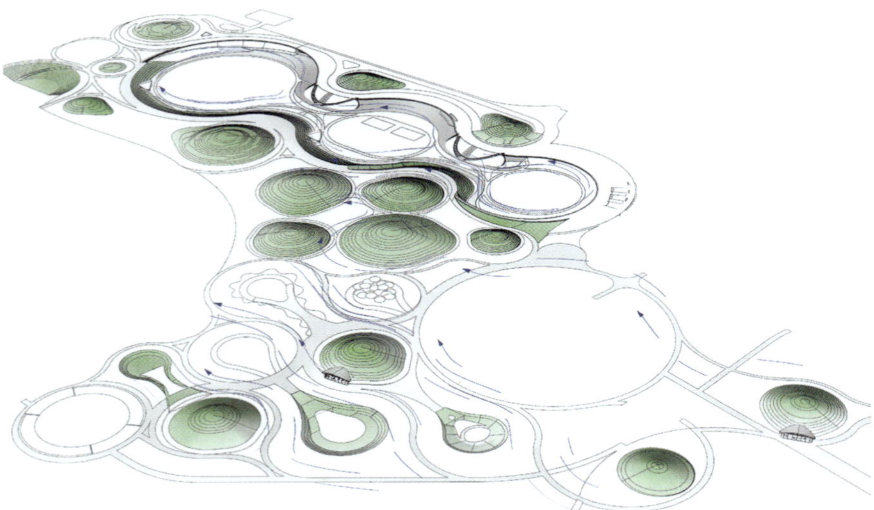

Tehran Entrance Plaza Station

Sadeghiyeh metro station was the first metro station built in Tehran. As a transportation hub, the design of the square by Mahdi Kamboozia and Helena Ghanbari echoes its function of intertwining different modes of transportation such as taxis, cars, and bicycles. The focal point of the structuring ripples was chosen by carefully studying the flow of pedestrians around the station. These ripples culminate in the main courtyard, which serves as the taxi station. Towards the east, approaching the entrance to the station, the ripples are interspersed with various functional constructions, such as benches, a place for merchants to display their products, and a green space. To the west is a space for public events. As a strong pedestrian axis cuts through the square, a space has been created where vendors and pedestrians can interact.

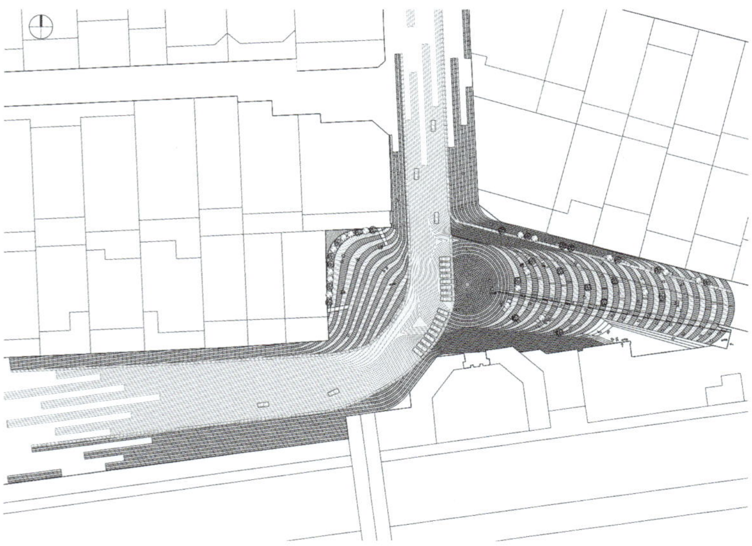

Northeast Pavilion at Surajkund Craft Fair 2023

The pavilion represented the northeastern states of India at the 2023 Surajkund Craft Fair. As this region is characterized by a high degree of ethnic diversity as well as indigenous forms of art and craft, the pavilion was used to allow artisans to demonstrate the workings of eight indigenous looms. The looms connect the space, which is divided into eight different courtyards. Each of the courtyards showcases a handloom from one of the eight northeastern states. The eight cylindrical galleries intersect the hyperbolic roof at different heights, reflecting the hilly topography of the Northeast. The central courtyard, representing Meghalaya, the wettest state in the world, faces the sky and contains a pool of water. The open and permeable design of the space blurs the boundaries between the pavilion and the surrounding nature.

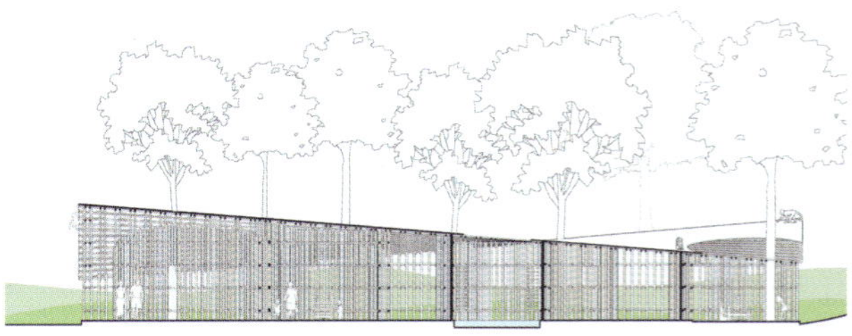

One Green Mile

One Green Mile is a paradigmatic example of urban renewal. It redefined a 1.8-kilometer stretch of Senapati Bapat Marg, aiming to craft a public realm that is safe, active, and vibrant, addressing the multifaceted needs of its diverse users within one of Mumbai's most hectic financial districts. StudioPOD designed a comprehensive urban regeneration master plan through a strategic public-private collaboration between the Brihanmumbai Municipal Corporation and Nucleus Office Parks, and the residential and commercial community along the identified stretch. Emplcying sophisticated road space reallocation strategies, including geometric corrections and road dieting techniques, StudioPOD succeeded in enhancing vehicular circulation and minimizing pedestrian-vehicular conflicts, thus prioritizing pedestrianization.

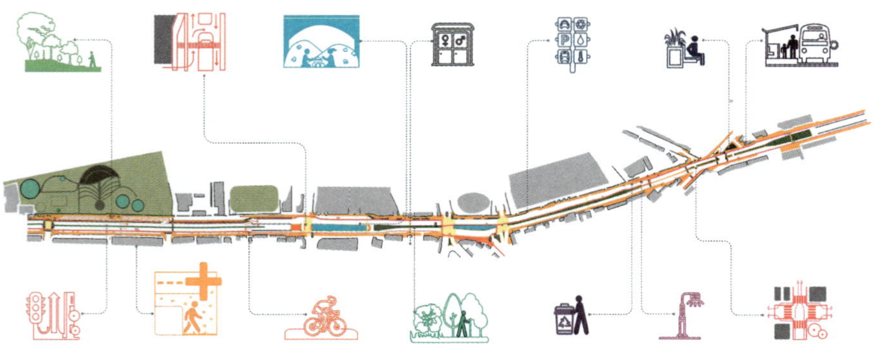

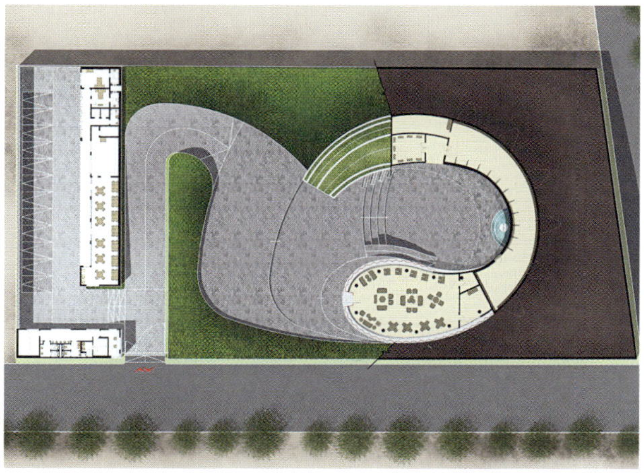

Nokha Village Community Center

Located in the Rajasthani desert, the Nokha Village Community Centre is designed by Sanjay Puri Architects. A north-facing open courtyard is created by the sweeping volume that ascends from the northeastern corner, loops around the site, and rises on the northwest corner. The brief of the clients was to create a memorial for their father. The 836-square-meter building transforms into an inclined garden with two different slopes, making it suitable for gatherings and recreational activities. From the rooftop, there are views of the surrounding desert terrain. Stone screens, the courtyard facing north, and a southern grass berm make the area energy efficient by decreasing heat. The structure and its construction are sustainable since rainwater is being collected and the labor and materials were locally sourced.

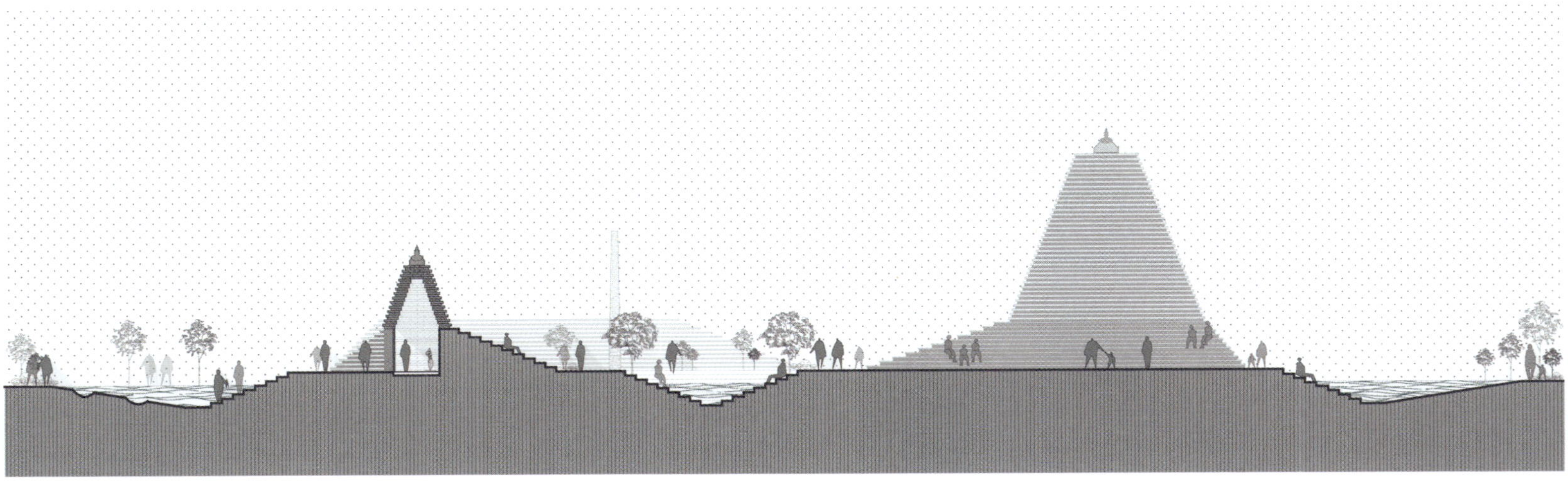

Balaji Temple – Temple of Steps

The Balaji temple at Nandyal explores and abstracts the long tradition of temple typology in India. The design of the temple itself is based on a 10th century temple for the same deity at Tirupati in southern India and similarly includes the Balaji & Varahaswamy shrines and a pushkarini (water tank). In the arid terrain of Nandyal, the main concern was to create a space that would reconcile the socio-cultural expectations of a temple with the ecological framework and dynamics of and around the site. Water overflowing from the limestone quarries was directed to a low-lying recharge pit, the banks of which were conceived as a social space in the manner of a traditional ghat; a flight of steps leading down to a body of water. This negotiation of land and water through steps is an important part of India's architectural heritage.

Kiss Bridge

The Kiss Bridge, born from a deep reflection on the concept of connection, consists of two large, curved arms that touch each other after a sloping section, without ever connecting out. Their physical distance of only 50 centimeters symbolizes the desire to meet. It recalls the image of the outstretched fingers in Michelangelo's The Creation of Adam and is inspired by the Vietnamese legend of Ông Ngâu Bà Ngâu. The legend tells of the love of a goddess and a man, so strong that the emperor allows them to meet once a year on a magical bridge created by a flock of birds. The goddess and the man will love each other without ever touching, a platonic union that resists forced distances. The work, evocative and symbolic, offers a place where different cultures and experiences can connect.

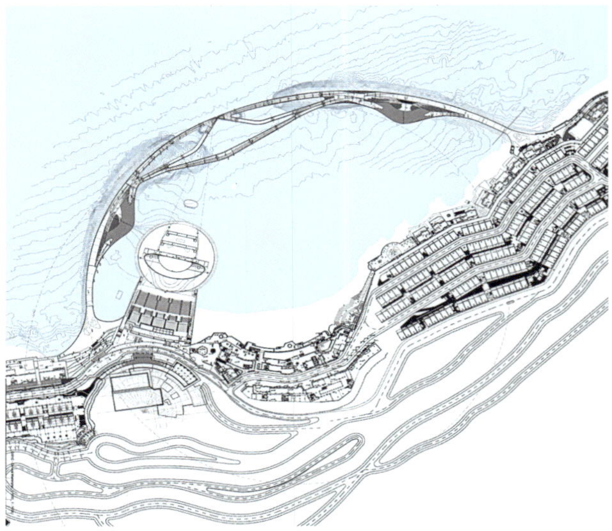

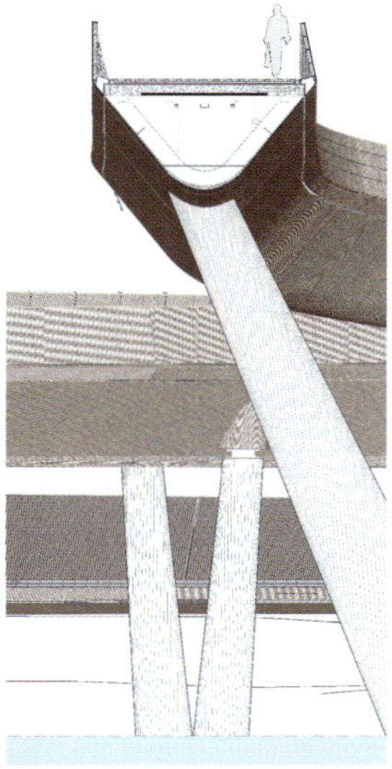

→ TK Studio → Education and Playgrounds → 2021 → Bangkok → Thailand → Photos: Rungkit Charoenwat, Weerapol Singnoi

Forest Pavilion, The Forestias

The Forestias is a large property development project in Thailand. The project aims to let people live in harmony with nature by providing them with a quiet place to find happiness. As part of this, a 48,000-square-meter urban forest has been created, where the Forest Pavilion has been built. In the grand scheme of The Forestias, the building served as the sales gallery and now serves as the Ecosystem Learning Center and gateway to The Forestias. The building embodies one of the central concepts of contemporary landscape architecture: symbiosis with nature. It achieves this through a green roof, tranquil trees, and a skywalk that connects to the large urban forest. The rigid form of the building blends in with, and yet contrasts with the free and rounded forms of nature.

Sierra Sathorn

The Sierra Sathorn project is an experimental prototype to promote the concept of the urban terrarium. In the fast-growing urbanization, the team plans to maximize the capacity of its landscape for new design standards of high-rise development. Sierra Sathorn considers not only the aesthetic aspect, but also the relationship between urban ecosystem services. The team is working with AR to transform a typical podium building into a new terraced landscape with green linkage space. In addition, the concept of this urban terrarium will create positive ripple effects on the future development of the private sector's green network.

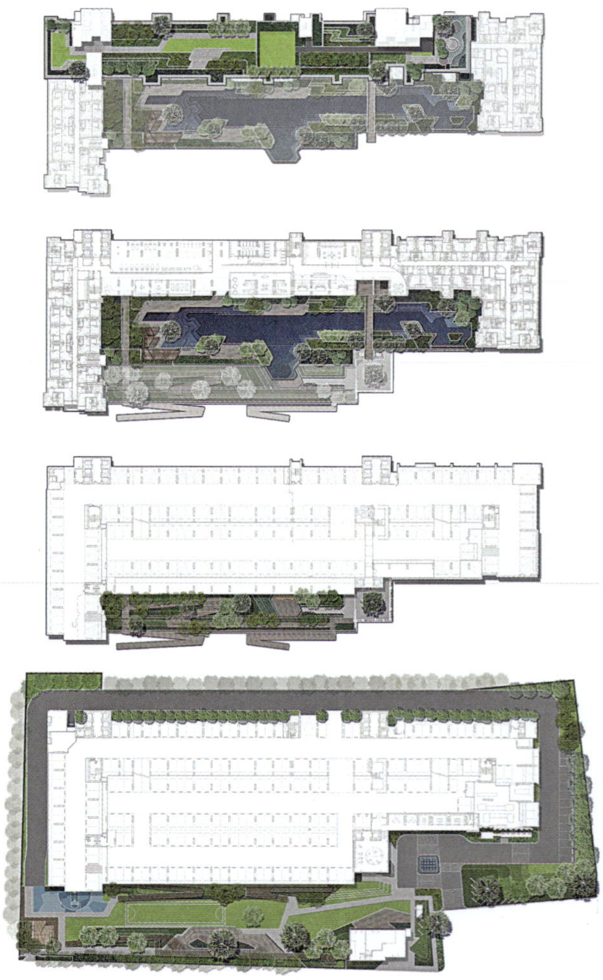

Chulalongkorn University Centenary Park

Chulalongkorn University Centenary Park is the first critical piece of green infrastructure in Bangkok to mitigate harmful ecological problems and reduce the risk of urban flooding disasters. By harnessing the power of gravity, the park is able to collect, treat, and store runoff. Situated at a three-degree angle, the park can hold up to nearly four million liters of water during heavy rainfall. The park is equipped with several ecological components – consisting of the green roof, wetlands, detention lawns, and retention pond – so that not a single drop of rain is wasted. The 1.3-kilometer-long road running perpendicular to the park, the first of its kind in Bangkok, has been redesigned to give priority to pedestrians and bicycles over cars.

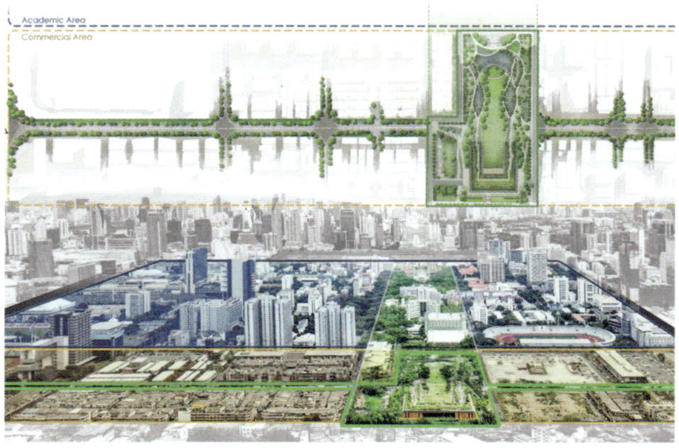

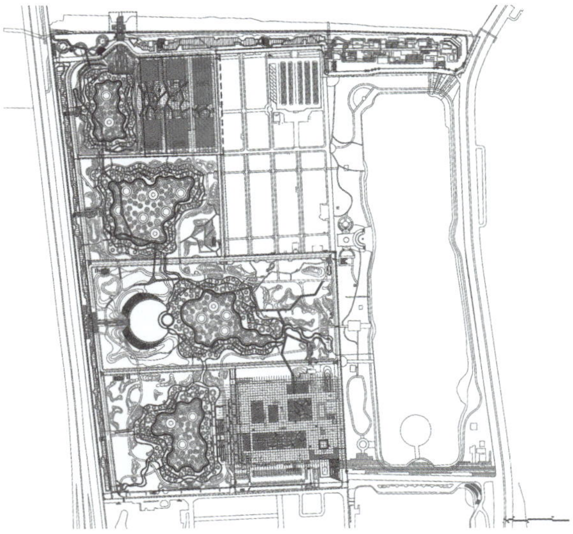

Benjakitti Forest Park

Benjakitti Forest Park, an urban oasis in Bangkok, was created on the site of a former tobacco factory. It covers 720,000 square meters and serves as an educational center on ecology, hydrology, and the environment. This initiative retains 128,000 cubic meters of rainwater, generating 1,600 cubic meters of clean water per day. It addresses floods and droughts with nature-based solutions and acts as a sponge, using bioremediation for purification. The park fosters a wildlife habitat, houses 360 species of trees, and transforms swampland into a rich ecosystem. It incorporates sustainable techniques, reusing materials and repurposing factory buildings for museums and sports centers. Accessible to all, it's a sanctuary within Bangkok that fosters a diverse, nature-centered experience – a biodiverse classroom for city dwellers.

Thanh Long Baywalk

Thanh Long Baywalk is located on the natural rocky shore that creates the shape of a cape and forms a breakwater point with sandy beaches and coconut forests. The square is carved by the spatial envelope character of the column corridor that integrates dragon fruit colored silk strips within the coconut forest. The structural columns resemble straight coconut trees, blending with the real trees thanks to their artificial bark made of fiber concrete. The white Moon Halo at the top of the Baywalk is made of metal cladding panels perforated to mimic the seeds of the dragon fruit. The ecological paving solution is provided using prefabricated, self-compacting concrete blocks on natural sand, facilitating the drainage of the soil. Visitors can walk on the sand, touch the green coconut leaves and capture the beauty of the bay.

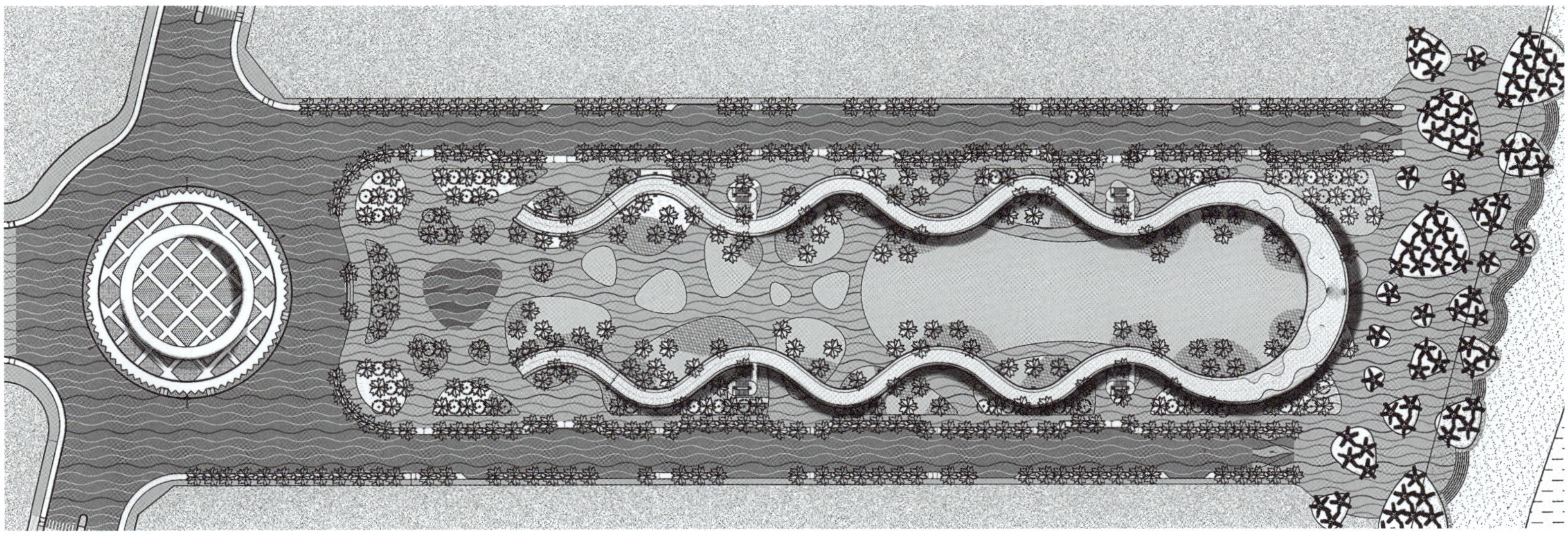

Mây Pha Lê – Crystal Cloud

Standing on Doi Mam Xoi plateau, surrounded by layer upon layer of rice terraces, and after observing the Hmong farmers tending the fields, Andy Cao und Xavier Perrot had a vision: This beautiful site could be the ideal location for their first temporary crystal cloud installation in Vietnam. They knew that creating such an ephemeral new artwork should be a close collaboration with the local Hmong community, highlighting the beauty of things handmade, and continuing the Hmong's long tradition of sculpting the land. The Crystal Cloud is made with hand-sculpted galvanized wire mesh, and adorned with 58,000 brilliant Swarovski Crystals. The cloud formation is supported by a series of randomly-placed galvanized steel posts, resembling a shimmering crystal forest.

→ Gustafson Porter + Bowman, ICN → Public Parks and Gardens → 2018 → Singapore → Singapore → Photos: ingenhoven associates / HGEsch.

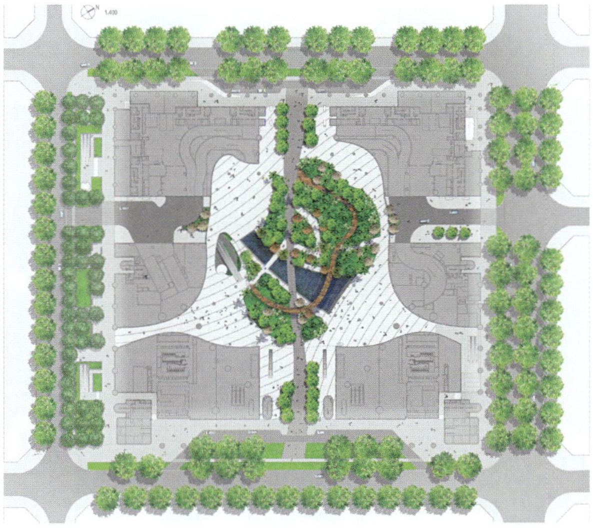

Marina One Green Heart

Marina One by Ingenhoven Architects supports Singapore's ambition to become a city in a garden. Gustafson Porter + Bowman's design for the central, Here, a Green Heart unifies the sculptural and planted elements of the building. While the outer face of the four towers strictly follows the city's grid structure, the inner space is a free-formed, three-dimensional garden that provides a rich and engaging experience for visitors. A winding ramp through the abundant vegetation creates both a visual and physical connection between the most important levels of retail, while water descends from above into a large reflecting pool. Openings between the high-rises, sky gardens, and elevated towers, as well as the shape itself, improve airflow, creating a comfortable microclimate within.

CapitaSpring

CapitaSpring is a 280-meter-tall high-rise oasis in Singapore's financial district, designed by CRA– Carlo Ratti Associati and BIG – Bjarke Ingels Group. The tower has Grade A office space, serviced dwelling, a hawker center, and public areas in addition to its sky-gardens and rooftop park. The building's vertical parts are pulled apart at multiple elevations to reveal greenery throughout. At the core of the building are four connected floors of organic softscape called the Green Oasis – a 35-meter open-air garden for work, casual strolls, leisure, exercise, and events. Weaving nature vertically into the architecture and satisfying the locale's spatial limitations, the Green Oasis gives tenants and residents abundant access to green spaces while blending modern architecture's elegance with the region's tropical nature.

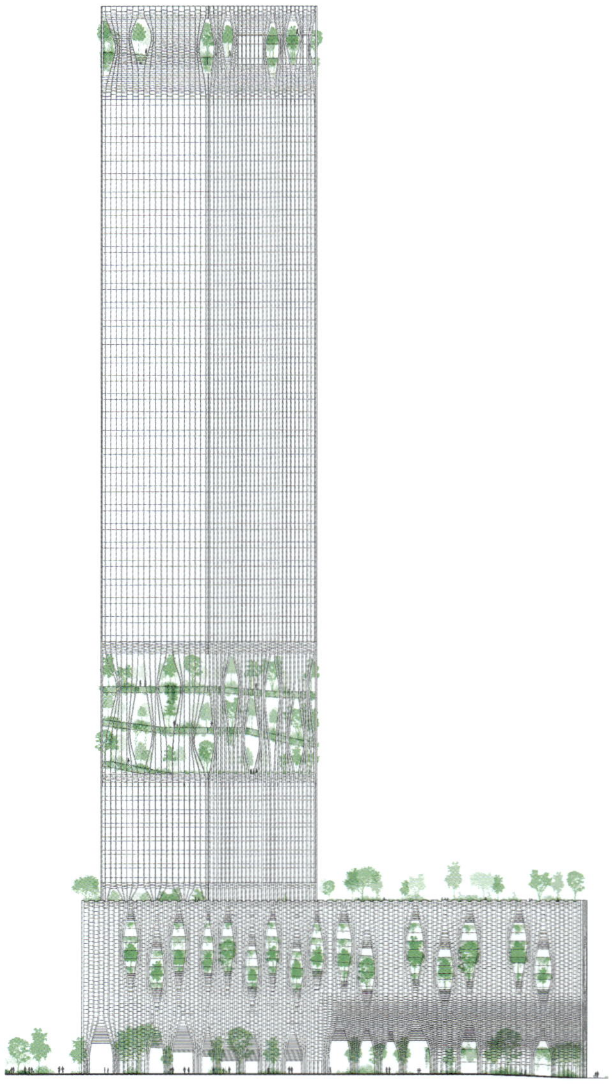

CapitaLand

NTU Yunnan Gardens

Yunnan Gardens, within the original Nanyang University, was a unique garden built in 1955 – a fusion of traditional Chinese, symmetrical French, and picturesque English styles. STX Landscape Architects in the rejuvenation retained key historic features of Yunnan Gardens, reinforcing its core essence, but updating it for the 21st century. STX created a series of bio-diverse inter-connected and inter-dependent spaces focused on community, with the historical layer still visible. Integrating 184 saved mature existing trees, a new river eliminates boundaries knitting together previously disjointed disparate spaces, as storm water management mitigation of run-off water from surrounding terrain. The new development consists of three main landscape design precincts: The Chinese Heritage Centre (CHC) Plaza, Nanyang Lake and Yunnan Gardens.

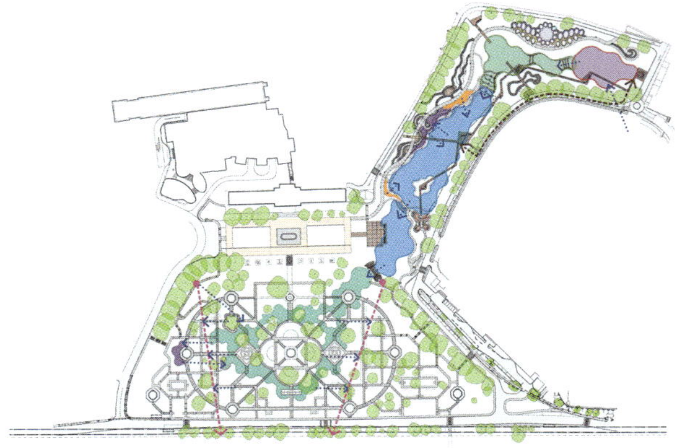

1955

Between Us 2

Two half detached houses are divided by a large open surface with timber decking. Four trees, protruding against a backdrop of lush landscaping, form the green spine of the interior. The formal symmetry of the floor plan reflects the idea of equality and balance. Hidden within the greenery is a large courtyard that brightens rooms in the basement. The greenery extends towards the rear in the form of a vertical garden. Hard materials, such as the walls of grey stone, battle with lightweight elements, such as timber screens, metal framing, and the thin flat roofs of the two main volumes and the car porch canopy. Closable timber screens enable the formation of a private area, as well as the regulation of the climate. This function accompanies formal articulations such as protruding roof eaves in crafting a sustainable tropical house.

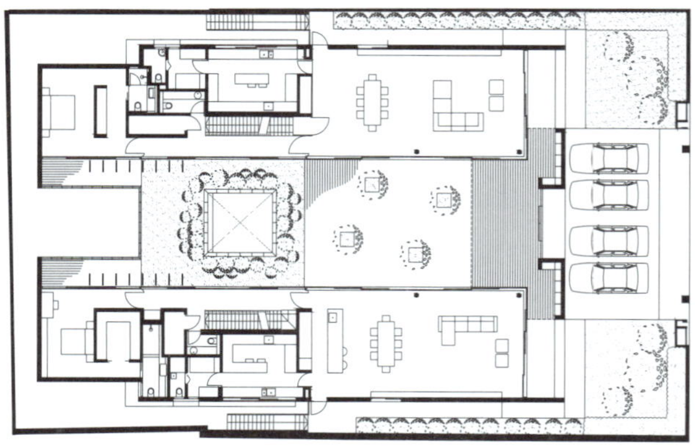

Hard Rock Cafe

Changi Terminal 2

The project spans three levels, including arrival and departure halls, and a mezzanine on level three where food and beverage options are available. Aiming to increase handling capacity, enhance facilities, and bring in more efficient systems new circulation patterns and welcoming environments were developed. The project was designed as a journey across an indoor landscape encompassing minerals, water features, and lush local vegetation in various forms and densities. Developed through parametric design, an intricately baffled ceiling composed of aluminum blades was created. It was designed to conceal every little detail, including sprinklers, security cameras, and floor lights that are tucked between the baffles, to remove the impression of an airport as being a cold and engineered space.

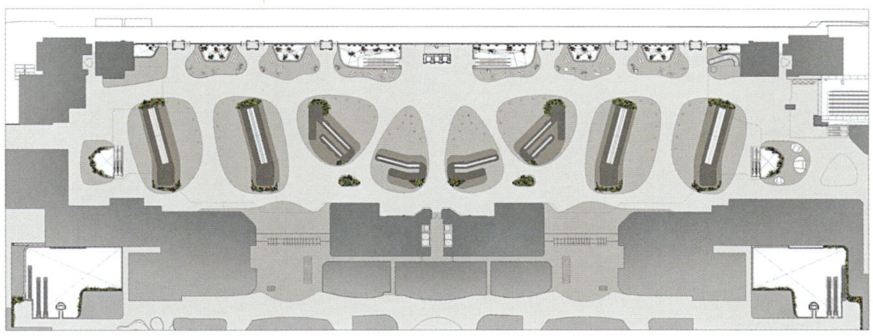

Nanhua Glimmer – A New Geometric Landscape under Taimu Mountain

The former campus of an elementary school in the rural area of Pingtung, Taiwan, situated at the foot of Taimu Mountain, has been transformed into the Pingtung AI Agriculture Hub, which is accompanied by an adjacent park. The landscaped park links its greenery with the surroundings. The reborn campus celebrates the regional nature features and translates them into geometric forms in the landscape design. The pattern of the concrete retaining walls evokes the shape of pineapple fields and the folds of mountain ridges. The selection of plants, with edible ones on the green hills and a restored eco pond with aquatic plants, provides an educational opportunity to the public. The irrigation system nourishing the plants, also produces mist, which evokes the image of fog flowing around a mountain.

Floating Park– Shinpu Elementary School

The project's aim was to realize a new urban campus core for a public school: a visually vibrant urban oasis with a greater variety of outdoor activities. The design concept reclaimed several building roofs, transforming ball fields into a multi-use activity space and redesigning spaces between the new and already existing school buildings. The typical stage platform was transformed from a rectangular base into a polygonal landscape element parallel to the active energy flow. A selectively chosen variety of compositional textures of individual design elements, different vertical heights, and color contrasts have successfully led to a dynamic redefinition of the activity areas. Green corridors, native vegetation thrives, planting beds, vegetative roofs, and special floor coverings are benefiting the new arrival of wildlife communities.

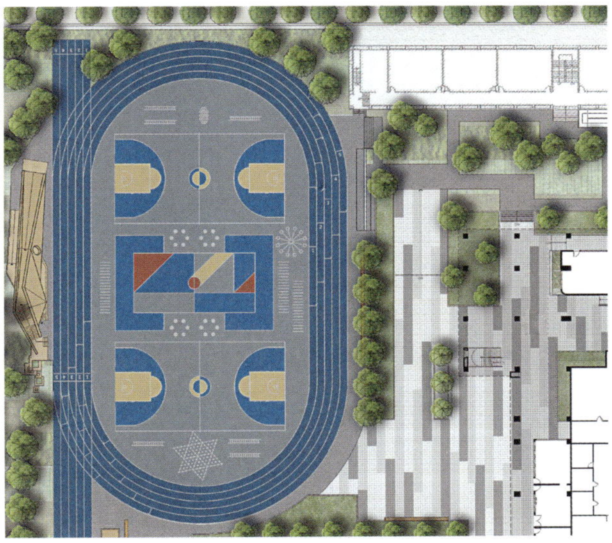

The Dark Line

As a hiking trail built through two abandoned railroad tunnels, this industrial site has become a hybrid environment where the classic distinctions between nature and culture are blurred. The trail preserves the post-industrial landscape in all its historical depth, while nature has produced ecological, atmospheric, acoustic, chromatic, and luminous variations over time. In this way, the intervention allows both the old structure and the landscape to speak for themselves. The iron elements, repeated ad infinitum, forms a transparent floor and evokes the memory of the former railroad that once ran through here. Looking through the floor, one can see the powerful action of nature. The steel bars gradually give shape to the ground, lifting and bending almost imperceptibly.

2018 Taichung World Flora Exposition

The topography of Taiwan boasts more than 200 mountains that soar over 3,000 meters high. Within Taichung's jurisdiction lie Snow Mountain, Dong-shi Forest, the urban metropolis, and the Kao-mei Wetland, showcasing a diverse range of topography, climate zones, and flora. Over 5,000 indigenous plants and 700 ferns contribute to the impressive array of species found in the region. The Forest Garden Area at the 2018 Taichung World Flora Exposition displays Taiwan's diverse terrain through a combination of agriculture and spatial aesthetics. The plantings mirror the growth patterns of plants at different altitudes, from the misty high hills to the broad-leaved evergreens of the middle mountains, to the banyan trees on the plains. This creates the most diverse miniature landscape in Taiwan.

Taichung Green Corridor

The 1.7-kilometer former railroad line through downtown Taichung was an important catalyst for the development of the old city. Later, the abandoned railroad became more of a barrier due to its challenging location on the dyke, making it difficult to cross from one side of the tracks to the other. The design of the new public park with herb garden, street fitness, tribune, playground, viewpoint, and water park reuses the railway line to connect the different parts of the city. Within this linear site, sustainable development encompassed a wide range of planning and design factors, including urban renewal, public participation, historic preservation, green and water resources, and circulation routes for pedestrians and bicycles.

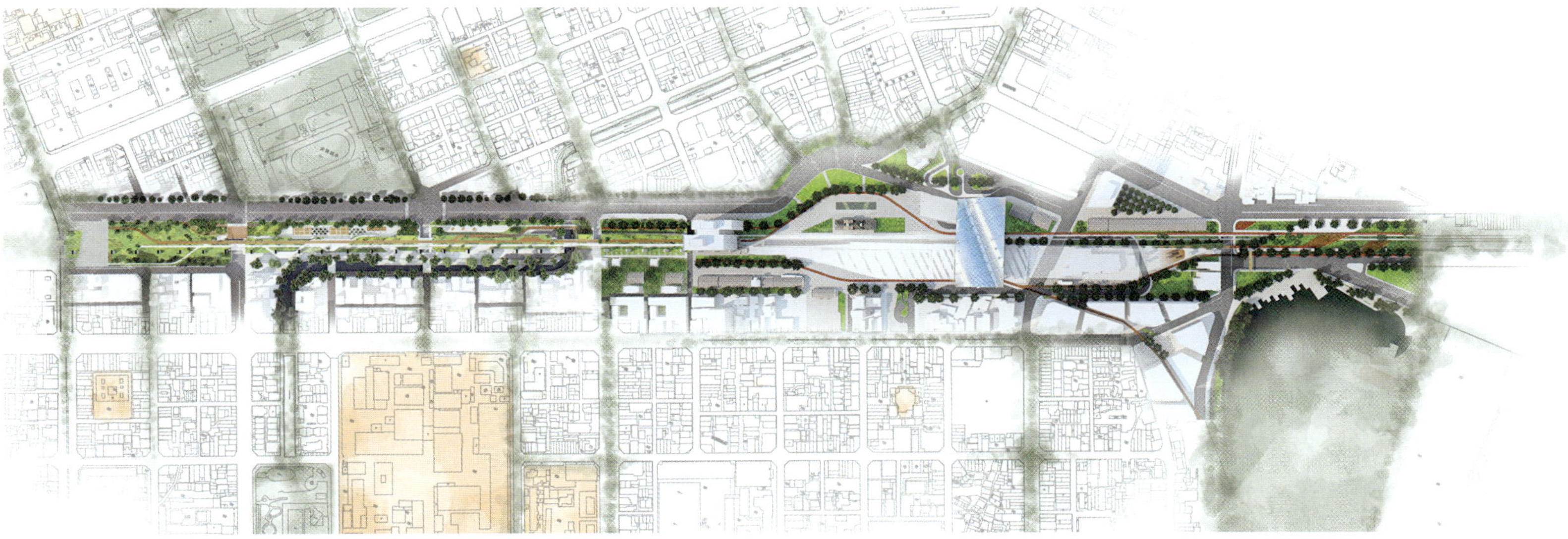

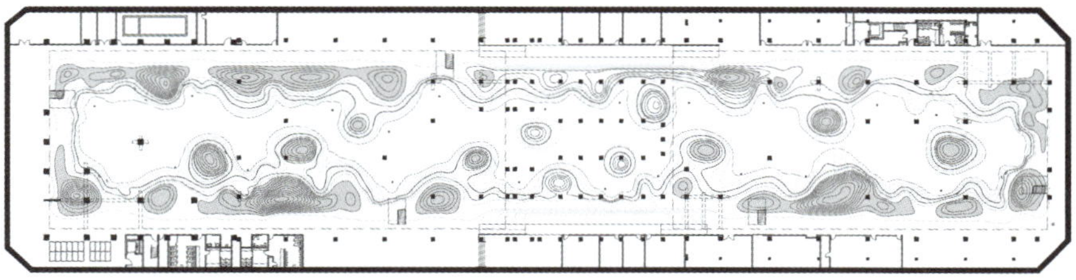

Tainan Spring

Tainan Spring is an urban oasis in the ruins of an abandoned shopping center. The underground parking deck has been transformed into a sunken public square dominated by an urban pool and native greenery, surrounded by a shaded pergola. The water level rises and falls depending on the rainy and dry seasons. Misting diffusers lower the temperature in hot weather to provide welcome relief for visitors and reduce the use of air conditioning in the summer months. In the remains of the carefully deconstructed and recycled concrete skeleton there is space for playgrounds, meeting rooms, and a stage, as well as kiosks and retail. Tainan Spring is part of MVRDV's master plan to green this part of the city. Together with the neighbouring Hainan Road, it contributes to the air conditioning of the city.

Taikang Water Dress

Since 2019, Lee, Kuei-Chih and Lua Rivera have been creating so-called Water Dresses that belong to unique fields in different waters. They let the work float on the water: Like a woven dress, using art to embrace the most important and mysterious substance – water. Salt making is very popular in Taiwan, so salt was the inspiration for the Taikang Water Dress. Salt making requires water and time, and in reverse water therefore is the medium of time. The concept of Lee, Kuei-Chih and Lua Rivera's work is taken from the linear landscape of the salt field. The project is a response to the geometric arrangement of the Tainan salt fields, which tangibly incorporates the water.

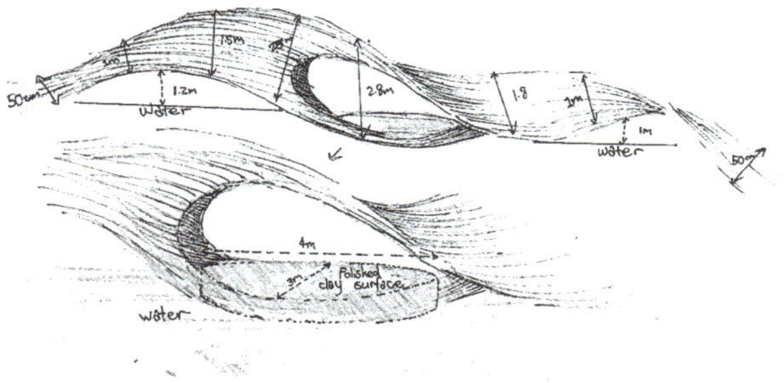

African Heartbeat

African Heartbeat is a garden that invites the visitor to experience the unique sensation of African hospitality. Surrounded by natural elements such as grass, wood and water, one feels at peace with only natural elements coming together to wrap their arms around you. Africa's flair for celebration comes to life in the decorative painting that depicts the diversity of life on the African plains. Water is the energy that connects the natural elements and allows for a sense of peace and contemplation throughout the garden. The sound of falling water adds to the wildness and evokes a peaceful African escape. Inspired by the African savannah, grasses become the main feature: Like water, they flow and move according to the gentle forces of nature. The festive spirit of the garden triumphs over everything, transcending borders.

The Crystal at Raffles City Chongqing

The largest indoor sky garden in the world, The Crystal is divided into three operating zones: hotel and dining facilities, the club pool, and the public observation deck. WAA's team procured each of The Crystal's over 120 mature trees to respond to the unique function and micro-climate of each space. The project's planting character was designed to celebrate Chongqing's rich natural environment, using plants procured from across southern China and acclimated locally to thrive within its climate-regulated interior, despite Chongqing's severely overcast weather. An innovative tree-replacement strategy was implemented by planting additional mature trees in on-site nurseries to be used as replacements for notable specimens, ensuring that the lifecycle – and legacy – of this suspended conservatory endures.

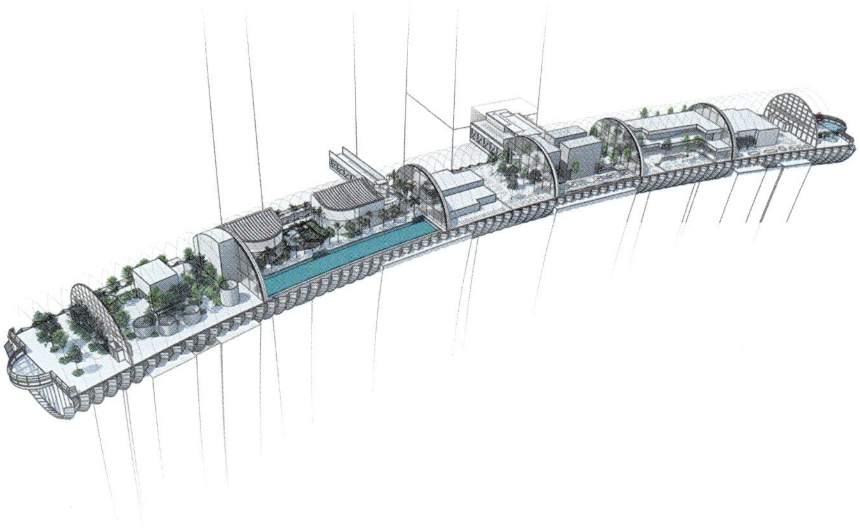

Chongqing Tongnan Dafo Temple Wetland Park

Tongnan Dafo Temple Wetland Park is located in the core area of the Tongnan City image exhibition area. Its adjacent to Fujiang River in the east and Dafo Temple. The site, whose periphery has been built for 20-year flood control, consists mainly of the shallow areas affected by flooding of the Fujaing River. A constructed 990,000-square-meter of wetland, numerous ponds, and various native vegetation communities are configured to form the matrix of wilderness. The designers used minimal intervention measures to preserve the geographical features as much as possible. From this base, a pedestrian trail system will be created to enhance the wetland experience. Dewdrop-shaped seating islands and metal pavilions invite visitors to linger. The original maple, poplar, and grass vegetation in the central river islands have been restored.

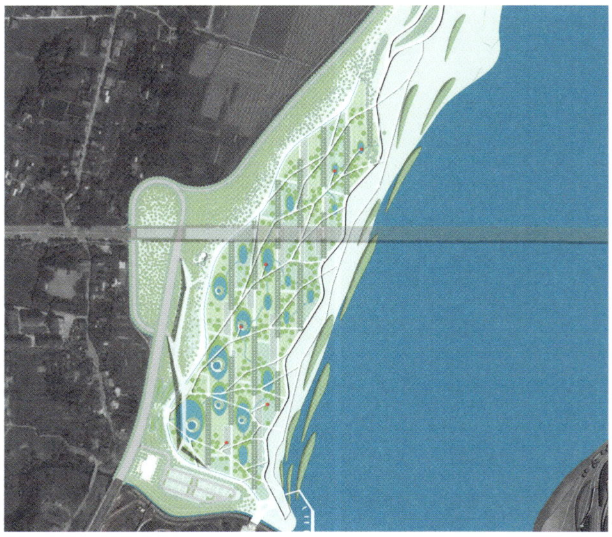

Blue Square

Blue Square is located in Dali Dong Village in the province of Guizhou, serving as an open space for communal gatherings and fests during important events. It was commissioned by The Earth's Tome Public Art Action and already hosted the first Dong Village feast during construction. Blue Square responds to the traditional Dong architectural principles. Four timbers clamped to cruciform steel nodes form the bundled columns, with very thin steel beams passing through the columns. The modern construction complements adjacent granaries. In the Dong village where indigo dyeing is a significant industry, the roof evokes the local drying rack – with hanging indigo-dyed fabric acting as a variable in the space. The canopy cover creates a sheltered space at a human scale for shade and rain protection.

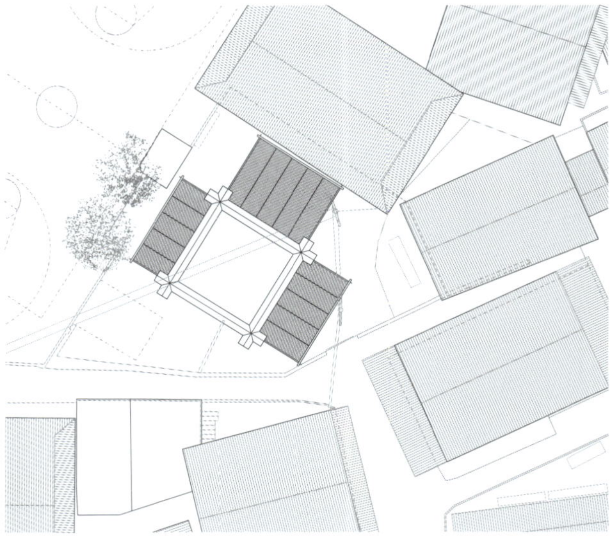

Natural and Organic: Red Dunes Playtopia

The Red Dunes Playtopia project designed by XISUI Design provides an attractive terrain for children to run, jump, and roll, referring to the theme of Mars exploration. The thoughtful integration of diverse children's activity installations with the undulating terrain creates a natural harmony that evokes the sense of a mountain nest or a jungle tree house. Utilizing advanced design techniques such as parameterized design, drainage optimization simulation, and structural form-finding algorithms, the Red Dunes project harmoniously blends the outdoor space with its natural surroundings. The overall topography design was generated digitally according to the complexity and slope of the terrain, providing distinct zones for different age groups.

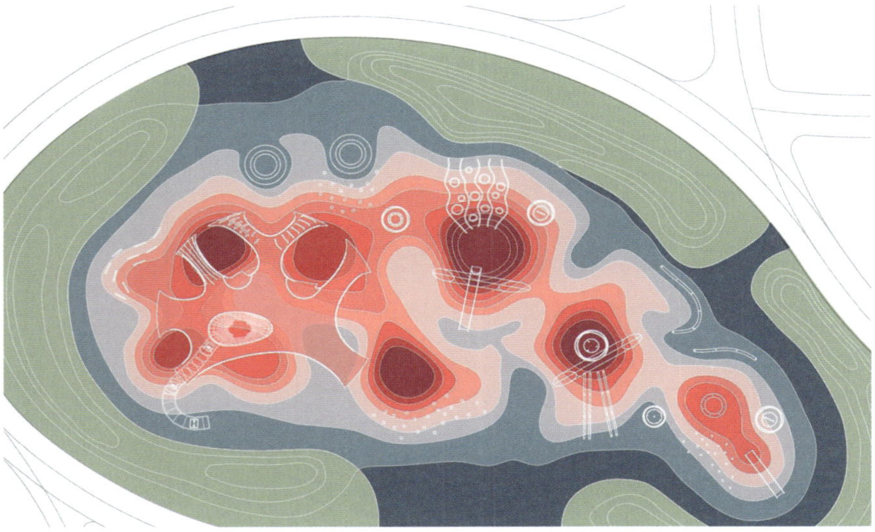

Hangzhou Green Town River Moon Residence

This project applies various design approaches to create a permeable landscape and architecture. It has used physical, visual, sound, and sensory elements to create a unique place distinguished by the penetrability of light, pedestrian flow, and its spaces' unification and flexibility. The façade creates big openings to the exterior space, providing circulation around the retail area. Garden seatings are introduced in the gray space underneath the architectural canopy, allowing a seamless connection from the interior space to the exterior space. Hangzhou is famous for its relaxing waterfront living. This public garden space demonstrates the lifestyle envisioned for this residential development. Water corridors and flowing plantings with pale green and blue palettes are featured this garden as a reference to the local culture.

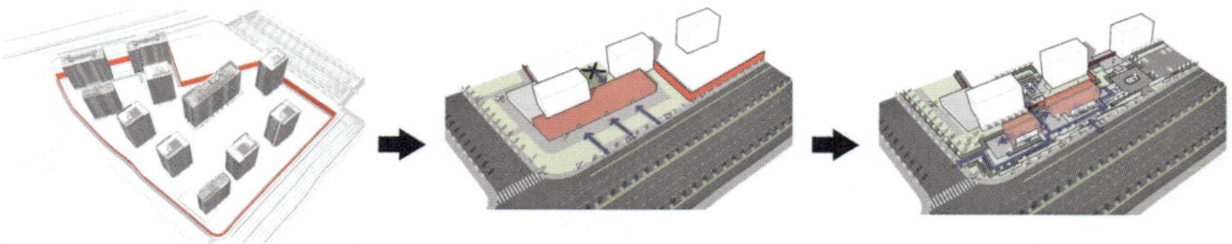

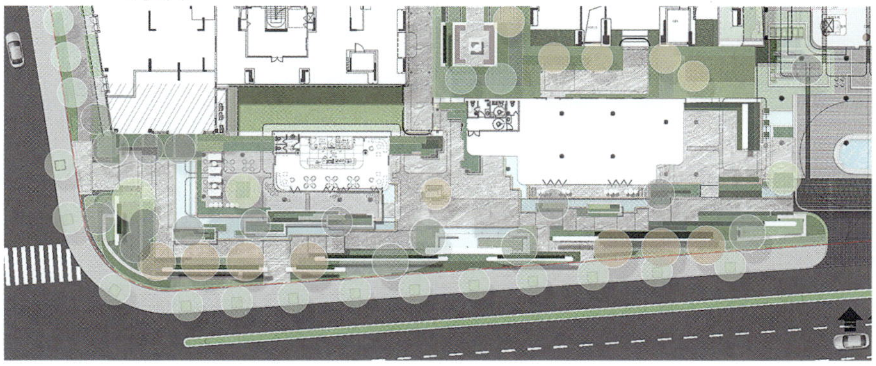

Lakeside Garden

Qiandao Lake, located in a densely forested valley on the outskirts of Hangzhou, China, is part of the Thousand Island Lake Scenic Area, which covers 448 square kilometers and is under ecological and scenic protection. Created by the construction of the Xin'an River Hydroelectric Dam in 1959, the lake is known for its clear waters, lush islands, forested mountains, and sublime natural beauty, including unpredictable floating fog and magical light. The site planning addresses the organization of commercial and residential uses, resulting in the framing of exterior public spaces and views. A meandering ascent reveals a series of symbolic gardens that use figurative landscape design and art to tell stories about this special place, including the lake's history of submerged villages and cities, and terrestrial natural systems.

Xiaohu Park

Located in Xiaoshan, Hangzhou, the project features an elongated and narrow valley and a traditional Chinese mountain-and-water landscape pattern. The name Xiaohu means small lake in English. The essential location of the site between the city, the local community, and the mountain-and-water landscape, requires a trigger of connection by minimal intervention to the sensitive ecosystem. The total area of the project is 60,000 square meters, whereby the loop around lake will fullfil a recreation function for future community development. While protecting the biological diversity of the site for ecological sustainability, the design team carefully laid out interactive facilities according to the different characteristics of the lake bank, marsh land, and woods, introducing diverse interactive distances between human beings and nature.

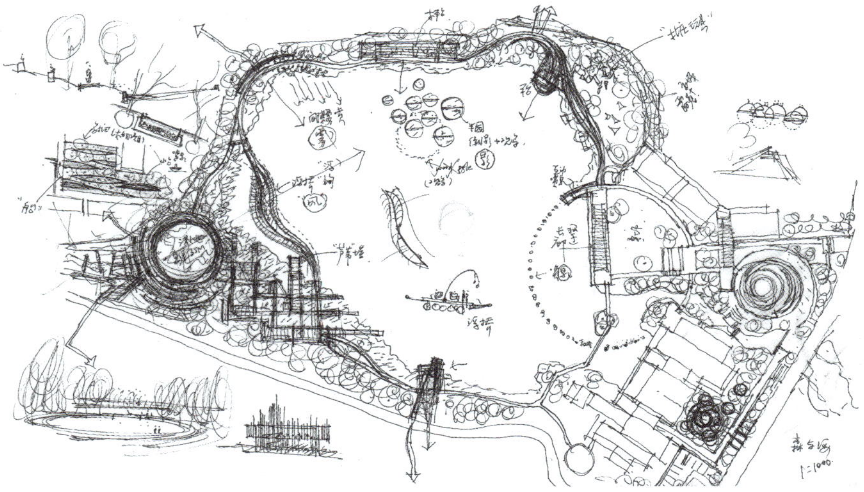

Silver Moon Bay

L'agence Belt Collins a conçu dans la ville chinoise d'Hefei un ensemble résidentiel de luxe, qui associe le respect séculaire des phénomènes naturels à des équipements modernes. Fondée sur des principes biophiles, cette réalisation favorise les échanges entre les résidents et l'environnement naturel, offrant la sérénité des retraites en montagne et la tranquillité des habitats en bord de lac. Les variations saisonnières du paysage ajoutent au plaisir des yeux, grâce au choix réfléchi des couleurs et des ambiances. Cet ensemble résidentiel est un sanctuaire où se mêlent passé et présent et où des éléments de la culture esthétique de l'Orient rendent hommage à la nature. En créant un refuge culturel et écologique, dans une évocation du patrimoine d'Hefei et du respect universel de la nature, l'agence a institué un nouvel espace de référence pour l'habitat de luxe en Chine.

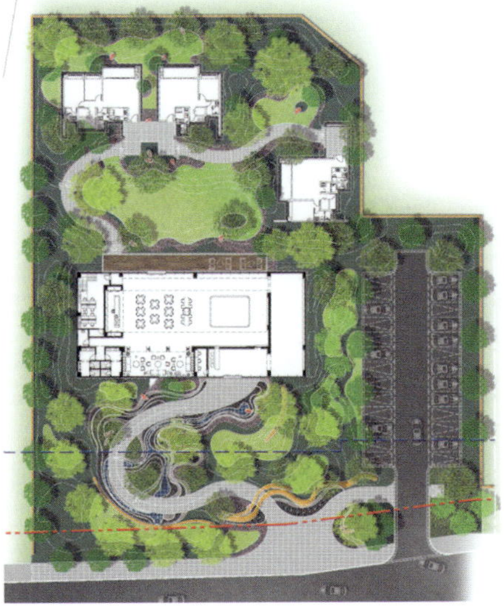

Miaojing River

The project transforms the Miaojing River corridor, a historically preserved waterway located in Kunshan West, into an elongated six-kilometer ecological infrastructure in form of a public linear park. The purpose of the project is to advance the development of climate-resilient, green infrastructure in Kunshan West, with the Miaojing River as the central axis of abundant ecological value, enhancing recreational programs for the surrounding communities. A bridge crosses the river to allow visitors to traverse it, encouraging walking in nature. A variety of trees and flowers create a picturesque feeling, while constantly changing the appearance and atmosphere of the park. As part of Kunshan West, the Miaojing River will become an invaluable natural resource for the future of the region.

Lijiang Hylla Alpine Garden

Within the Pan-Himalayan region, Alpine Garden is located on the slopes of Jade Dragon Snow Mountain at an altitude of 2,560 meters. The design effort is based on the site, establishing a distinct closeness with the indigenous Naxi culture and alpine nature. The inspiring use of native plants, snowmelt, local stone, and ethnic textures not only creates a variety of spaces and views, but also draws visitors in to explore the rich local culture and truly feel that they are not just a visitor, but a local who belongs here. As a slice of modernism in a rustic setting, Alpine Garden challenges the notion of contemporary design in a rugged alpine location surrounded by magnificent scenery. By blending ancient Naxi and contemporary design, it serves as a strong educational landscape place to demonstrate the importance of Naxi gardening tradition.

Sky Bridge HQ

Sky Bridge HQ is located in the hub area of Hongqiao Airport in Shanghai. It was formerly developed as Sky Soho designed by Zaha Hadid Architects. The upgraded original sunken plaza had to retain the iconic and signature forms of the original ZHA. To do this, TOA Landscape Architecture adopted the design of flexible and movable landscape units that are prefabricated and have a short installation time on site. There are several modular units such as round bench, crescent bench, table and planter in different sizes. Trellises with climbing plants bring greenery to the plaza and provide shade. Originally, there was a huge waterfall from the ground level to the sunken level at the end of the plaza. TOA transformed it into a multi-way water feature and event stage while maintaining the pond's shape and structure.

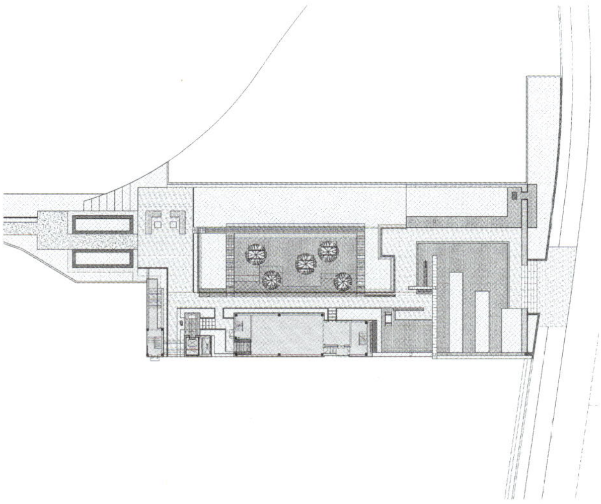

An Villa

Designed as an ode to old Shaoxing city, the project An Villa uses its internal landscape to manifest nostalgia and memories of rain in its architecture. The theme of rain is significant, as the region is known for its heavy rainfall and the city for its network of canals, bridges, and picturesque alleys. The black color of the curved shapes is interspersed with the green of the plants and trees. Different types of plants, trees, and grasses create a unique atmosphere, harmonizing perfectly with the surrounding architecture. The project combines nature, color, and form. The result is something that resembles a traditional memorial roof, but is at the same time transforming into a new architectural motif that has never been seen before.

Shenzhen Coastal Primary School

The landscape design concept of Shenzhen Costal Primary School – architecture by Shenzhen Architecture Design Institute – is creating a unique outdoor environment for both leisure activities and education opportunities in a school campus by blurring the typical boundary between the learning-oriented buildings and the outside space for playing. The design on 13,400 square meters forms the landscape in a pixel language that is consistent and unifying multiple levels of the school, which encourages learning through play while being inspired by the landscape. As many additional schools are planned to open over the next few years in Shenzhen, an immigrants city of China, this project hopes to inspire designs that embody a natural, engaging, and safe environment that students enjoy.

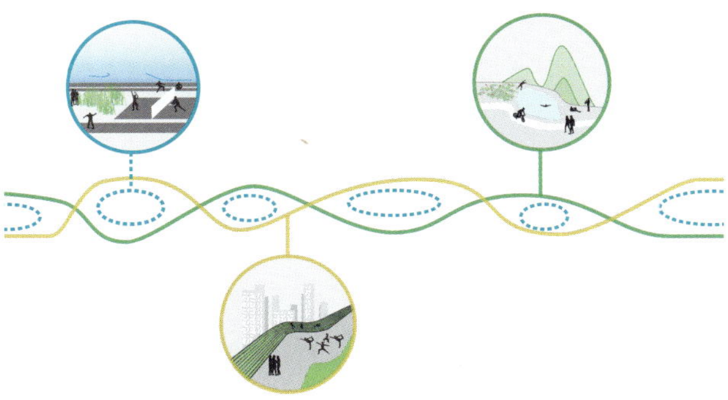

Shenzhen Skypark

The Shenzhen Skypark project revalues a former unused rooftop of an infrastructure building, converting it into a recreational strip, with the Shenzhen skyline as a stunning backdrop. As a way of giving back to the neighborhood, the park connects to adjacent buildings via bridges, bringing students and sports enthusiasts on a platform designed for them –connecting educational institutions with the local community. The unique combination of sports and recreational offers are being tied together with the landscape. The park aims to enable a healthy urban lifestyle that complements the city's youthful, entrepreneurial spirit. The positive atmosphere has an encouraging effect on those doing sports here, and Shenzhen Skypark is now a cherished place where many people gather together.

Suzhou Longfor Paradise Walk

The project is a retail development divided by a waterway, with a larger west block, a long and narrow east block, serving as a children's activity center, and a central courtyard plaza facing towards it. It is a community-based commercial complex located on the banks of Yangcheng Lake. The region is famous for a long-standing culture of crab-eating during the Chinese Mooncake Festival. The moon becomes the theme, seeking to express a dynamic landscape that appears to be constantly changing, just like the lunar phases. Topographical mounds were created for staged play with an integrated playscape. The design is built upon ensuring inclusion and diversity through considering local traditions, while also integrating a series of multi-functional spaces that encourage people to socialize, play, observe, and rest.

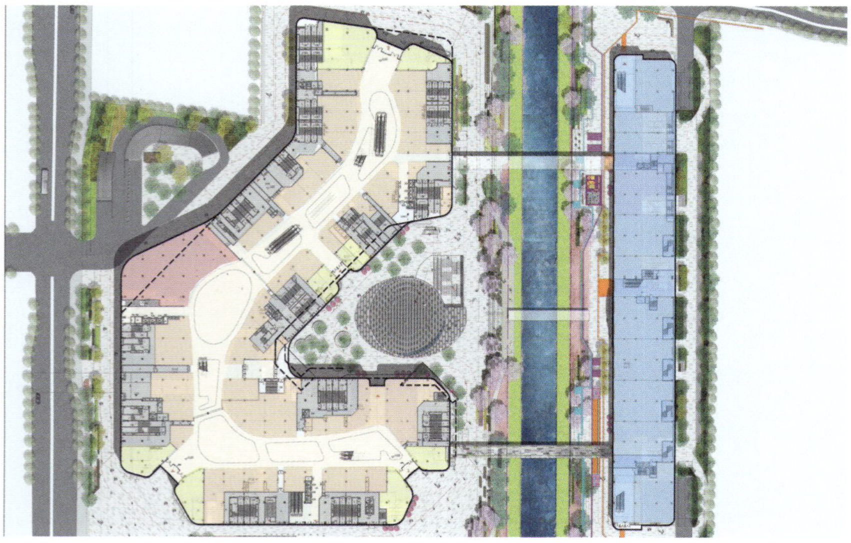

Paradise Walk

Bamboo Hill – Tianjin Vision Hill Plaza

Wisdom Hill Park is located in the hi-tech district of Tianjin, where the architects wanted to create a space that is culturally attractive for the urban environment. The main bodies of the hills are constructed a combination of bamboo slats, bamboo beams, and steel joints. Their open-flow structure contrasts the sharp structures of the high-rise concrete buildings that surround the area. The hills practically offer a landscape suitable for all kinds of open activities. As a living material, bamboo has a unique cultural and emotional connection to Asia, whilst its usage in this type of environment represents the symbiosis of rural nature and the metropolis.

North Pavilion in Wuhan

Wuhan North Pavilion is a 65,241-square-meter retail space that explores the quality and treatment of nature within a commercial space. A large atrium and sky bridge anchor the center and at the central court, a tree column roots the two-level interior in a thematic expression of nature and organic design. The exterior offers tree-lined terraces ascending from the street level, pulling visitors onto the outdoor platform. A rooftop garden is accessible from the ground level and opened to the public. The North Pavilion garden spans the entire length of the mall and showcases the true integration of public space. Seasonal flora and a variety of tall grasses are mixed to create an ecological haven that pays homage to the regions horticultural past and introduces a design concept that evokes Wuhan's pastoral landscapes.

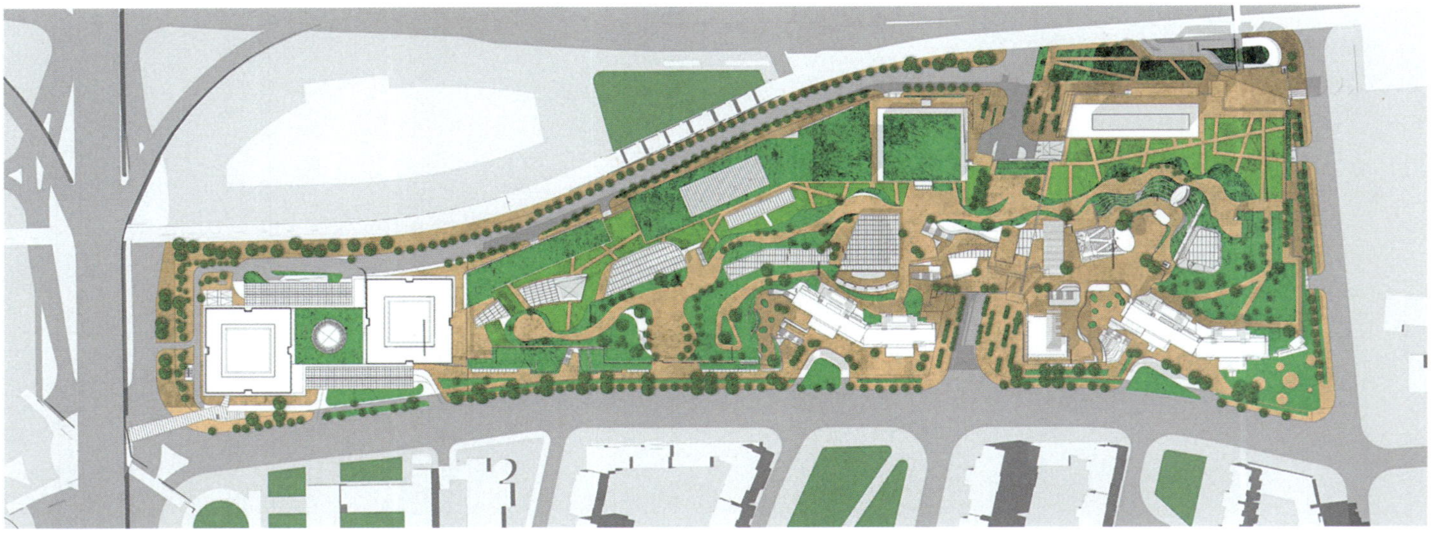

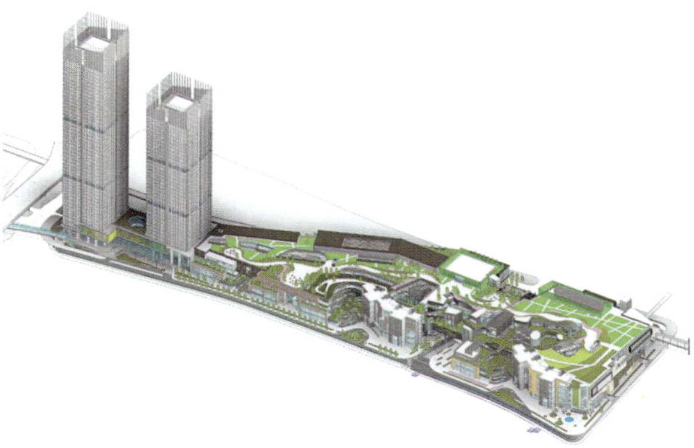

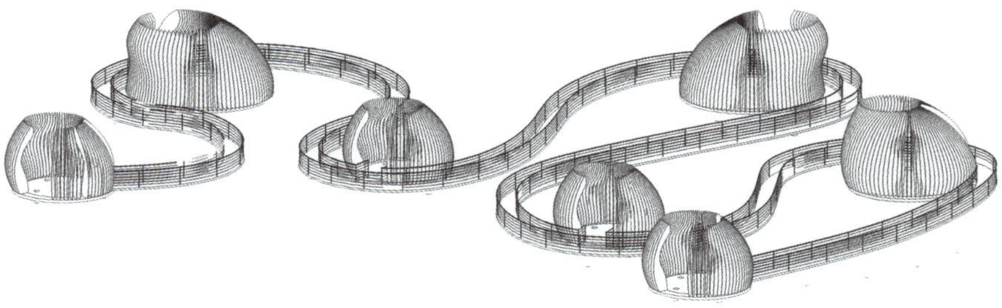

Zhuhai Garden

This project was designed as a show garden to represent the city of Zhuhai at the 11th China International Garden Expo. Zhuhai, whose name can be literally interpreted as "The Sea of Pearls", is a modern coastal city with a curving coastline, numerous islands, beautiful beaches, and lush trees. Inspired by these features, the design focuses on the theme of the sea and pearls: a curving coastline, a pristine white shell beach, a series of islands, and a meandering bridge structure connecting six pearls. The trail varies in elevation and offers different perspectives. The pearls serve as pavilions for visitors to relax and enjoy the view. The planting is predominantly green foliage trees with minimal flowers, accentuating the prominence of the white pearls and beaches, while avoiding a colorful background.

Zigong Dongxingsi Waterfront

The project's landscape design reconnects the city to the waterfront, with multiple new amenity spaces and a strong visual identity, inspired by Zigong's history and culture. The steep slopes between Chongqing and Chengdu, which lie along Zigong's polluted Fuxi River, were home to dilapidated houses and suffered from recurring flooding. Now a sponge park in the adjacent residential development absorbs, filters, and attenuates stormwater. The landscape concept is inspired by the dragon, the spiritual symbol of China, which is celebrated at the local dragon boat races. The sinuous design motif reflects the water swirls of the meandering river. Striking sculptural elements in the form of dinosaur ribs serve as support structures for the terraces and reference to vast numbers of Middle Jurassic dinosaurs found in this region.

Magok Central Plaza

The site is located in the heart of Magok New City, connecting Magoknaru Station on Line 9, Line 5, and on the Airport Railroad. A 60-meter-diameter sunken square was built in the middle of the three stations, connecting the ground and below-ground levels along the exits of the station. The shape of the subway entrance is streamlined to correspond to the surrounding landscape and give a sense of unity with the curve of the square. The square has been left as empty as possible to accommodate various activities such as events, performances, and exhibitions. There is greenery to the southwest of the site and Seoul Botanical Garden to the northeast. Additionally, it is surrounded by hospitals, shopping malls, offices, and research centers which support the nearby residential complexes.

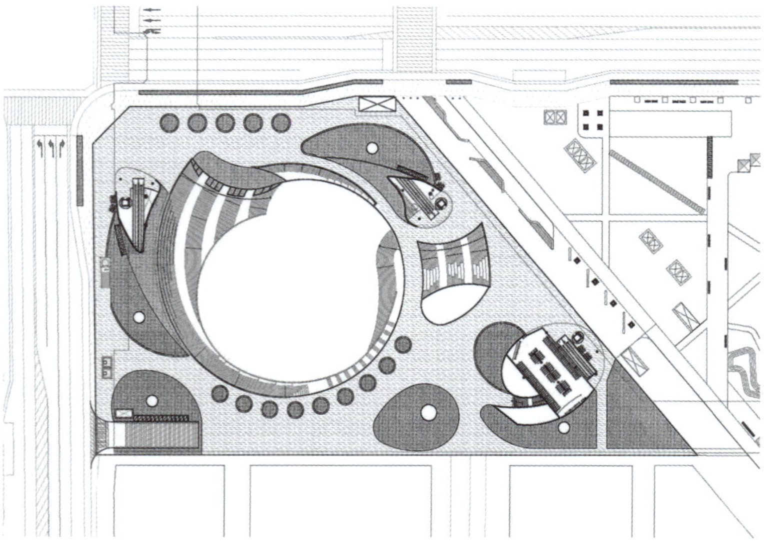

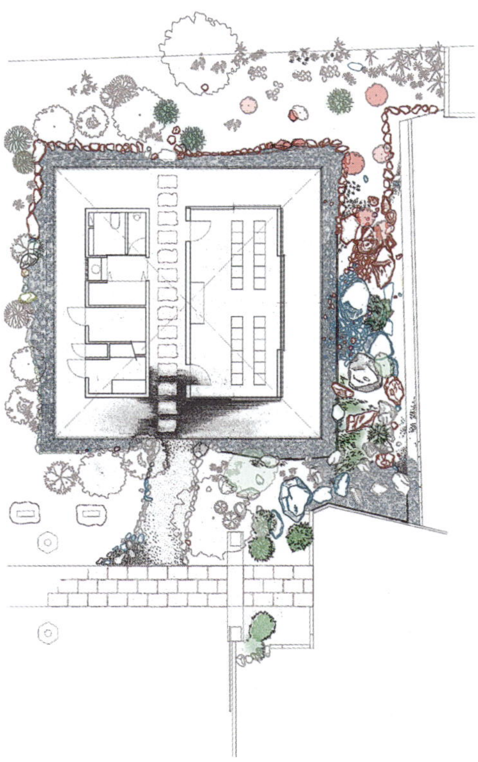

Tricolor Garden

This stone garden surrounds the newly built worship hall by Love Architecture in Saitama Prefecture. The project was a renewal for the entire temple, envisioning the next hundreds of years for future generations. Huge stones "Without a Name" take the lead role and were intentionally structured, some placed at an angle at the temple gates to express dynamism, some placed majestically near the building, and others placed with a certain serenity. One red and two blue stones are for zazen (sitting meditation). The front view gives a dynamic impression, while from the inside, the interior feels integrated with the ground. The new plantation of maple, yew, peony, and hydrangea has been divided into zones, and in autumn the colors of the leaves and stones turn the whole area red.

Kakumei-Kyukoh Garden, Reigen-in

Reigen-in is one sub-temple on Kennin-ji grounds. So, the Kakumei-Kyukoh Garden is a new garden in an historic Zen Buddhist temple in Kyoto. The dry landscape in this Karesansui style garden expresses the spread of Buddhism and Zen teachings from India to China and then to Japan. This is depicted in the L-shaped garden space to the east and south like in a large picture scroll. The eastern garden, which expresses the history of India and China, is composed of sweet hydrangea hills with yellow stones delivered from India, representing the birth of Buddha in India. The sacred Chinese mountain of Buddhism, Mount Songshan, is symbolically expressed by a dynamic rock arrangement. The South Garden tells the story of Japan with a few stones carefully arranged to symbolize the prosperity of the temple and its descendants.

Nara Hotel

The Nara Hotel was designed and completed 115 years ago by Kingo Tatsuno, who designed Tokyo Station, on the hill of Daijyoin Garden. Now the gardens have been redesigned on a former parking lot in front of the hotel. The entrance garden consists of a Karesansui (a dry garden) with a stone waterfall and weeping cherry trees planted on a Tsukiyama (an artificial hill). The garden on the right has a hill opposite with a stream in the middle. This design is reminiscent of gardens depicted in the Heian period scrolls. When it rains, rainwater flows through the grounds. The large roof of the Nara Hotel and the thickness of the existing downpipe made it possible to create such a Niwatazumi (a natural stream) here too. The most difficult part of the construction was to build a hill and plant trees in a limited space.

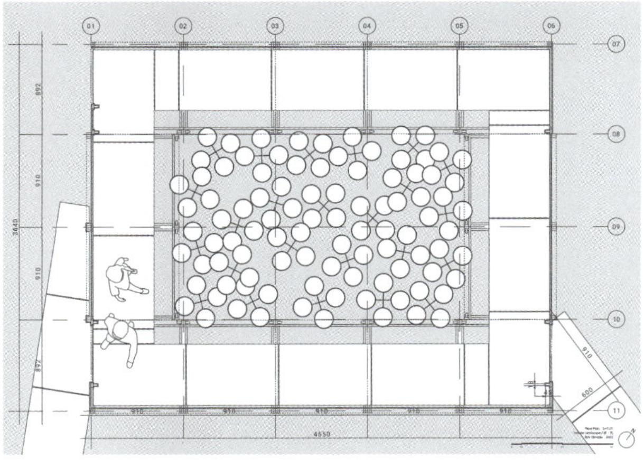

Infinite Landscape

This is an installation work of art set up to the center of a lake. The architecture, called the Corridor, is a 3.6 meter wide, 4.5 meter long, two meter high angled space. Visitors enter the Corridor through a passageway that is 27 meters long in total. The water center of the Corridor consists of one hundred uniquely shaped "Mirror Water Lily"-objects beautifully arranged on the water's surface. Each lily is 200 millimeters in diameter and 200 millimeters in height. It consists of an one-millimeter-thick iron plate welded in an on a vibratable steel rod that is six millimeters in diameter. Visitors can only see the color of water and sun rays and reflections. The feel of the corridor changes depending on the weather and time of day.

Zen Shrine in the Hotel Chinzanso Garden

The Hotel Chinsanzo is located in the center of Tokyo on Tsubakiyama, a hill covered with camellia flowers. In the Meiji Era (1868-1912), Japanese Prime Minister Yamagata Aritomo assigned Iwamoto Katsugoro, Tokyo's leading gardener, to create a Kaiyuu-style garden, which was later decorated with historical monuments from all over Japan. In 2017, the Mori no miya Garden Sanctuary was built; a shrine that combines Shinto shrine elements with contemporary architecture. 1Moku architects created a rooftop garden planted with camellias, the symbolic flowers that gave the mountain its name. The existing trees were used in the design of the green garden, so that the shrine blends in perfectly with the park of the Hotel Chinzanso.

Genbudo Park

The redevelopment plan for Genbudo Park in Toyooka City focused on improving the visitor experience at this nature park known for its columnar joints. The columnar joints are geological features that formed about 1.6 million years ago when magma cooled and solidified. Key improvements for the park's redevelopment plan included the creation of concrete platforms in front of caves with large columnar joints, providing a place for visitors to appreciate nature. Issues of color, fencing obstructing views, and existing structures were addressed by changing wall colors, redesigning fencing, and optimizing facilities such as the ticket booth. The goal was not a complete overhaul, but a strategic re-edit to maintain the park's charm while updating problem areas.

Cicendo Steel Park

The park features open-air sculptures as well as spaces for numerous activities, such as a skating area, a basketball field, an art market, a water playground for children, and street vendor kiosks. The park uses contour to differentiate between public and private areas. Weathered steel ribbons are used to organize the flow of visitors around these functions in the form of a topography of stepped hills that lend its form to the surrounding landscape of Bandung. Steps around the main square create spaces of rest and movement that encourage visitors to explore and take different vantage points along the park. The top of each hill, resembling pedestals in museums, serves as a platform for local artists to display steel sculptures. Behind the street vendor kiosks, there's a deck which serves as a quiet area, overlooking the whole park.

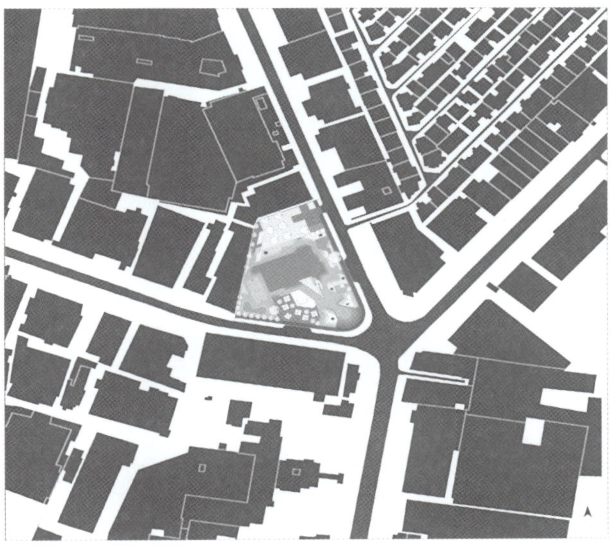

Tebet Eco Park

The Tebet Eco Park project revitalized a 73,000-square-meter public park in Jakarta, Indonesia within 15 months. The park was divided by a 714-meter-long open sewer that was heavily polluted due to unfiltered water runoff from the upstream catchment area. The ecologically restored river, planted with riparian vegetation, improves water quality by filtering and purifying run-off water from the surrounding area, increasing the biodiversity of the site, and reducing the risk of flooding. Excavated materials, such as existing stones of the canal and felled tree trunks, were reused and recycled as part of the new park construction material. The focus is on actively regenerating the ecology of the site by preserving trees and improving the blue-green infrastructure, as well as creating an open space for people of all ages and backgrounds.

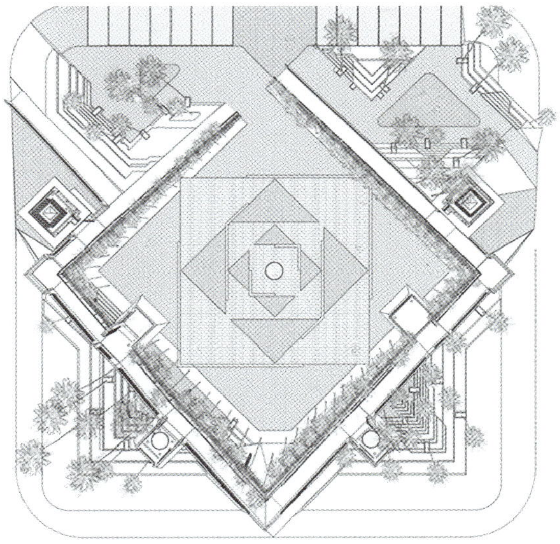

Octagram Plaza

Pamulang University chose Antonius Richard and RAD+ar to design its new Islamic-oriented public square, the plaza building and public artwork. The Octagram Plaza shows spatial gimmicks in its three-dimensional planning by using the important Islamic symbol Rub-ei-Hizb, which has the form of an Octagram consisting of two intersecting squares. This facilitates the recitation of the Koran in the garden, while the new community building and plaza are based on the concept of loop, connectivity, and permeability. The project combines hybrid contemporary tropical architectural and landscape approach with the heritage of the city. It has an outdoor gallery, a hidden amphitheater, a floating discussion podium and a garden terrace arranged in a flexible, permeable and free-flowing space to encourage the exchange of ideas.

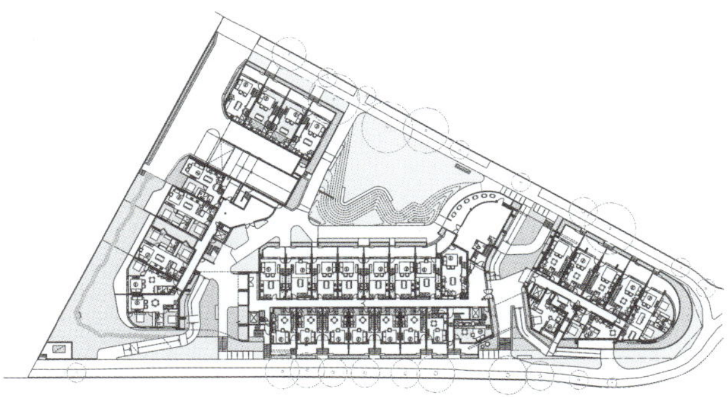

Arkadia

The project began with interviews with the surrounding neighbors and a historical study of the working-class community that existed there in the 20th century. The four buildings built to house defense personnel and their families, as well as renters from the open market, are joined together to form a single sinuous skin, woven from carefully crafted recycled brickwork. The building skin is designed not only as a reminder of the site's industrial history, but also as a beacon for a low-carbon future. Built with half a million recycled bricks, the building envelope is near carbon neutral and runs on 100 percent of electricity. Each building has its own communal garden and terrace planted with indigenous species, as well as a communal green roof. Arkadia thus combines environmental and social sustainability.

Barangaroo House

Barangaroo House, designed by Collins Turner Architects, the public domain landscape by Aspect and Oculus, and the interior design by H&E Architects and Studio Etic, stands as a bold statement piece in a backdrop of a highly urbanized skyline. The curved timber façade effortlessly wraps around the building, creating a bold cantilevered circular form. A series of custom modular balustrade height planters curve around the entire perimeter of these external balconies on each level. The mixed plantings, installed and planted by Junglefy, spill over profusely in an untamed form as they cascade down multiple levels, creating the illusion of an untouched landscape. The green curtains formed are most prominent along the east facing city façade, where the tall surrounding buildings create a low light microclimate.

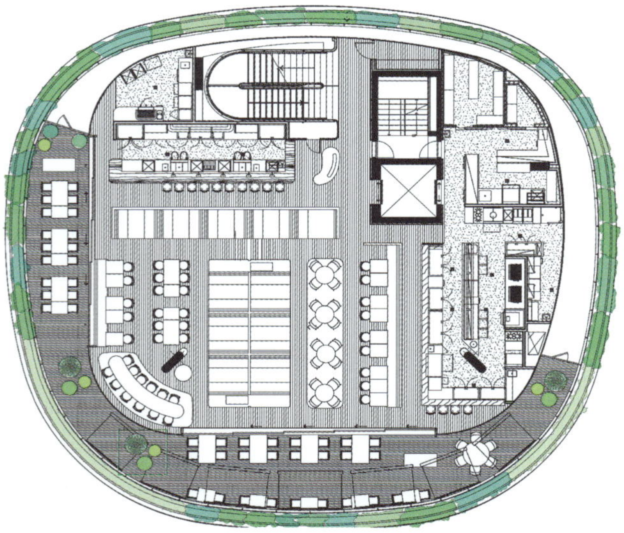

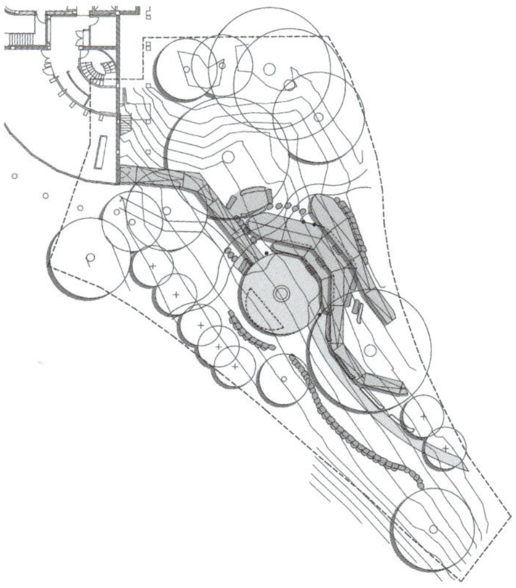

Yungkurrinthi Inparrila Cultural Gathering Space

Yungkurrinthi Inparrila, meaning meeting place in Kaurna, is designed as a place of cultural knowledge, learning, and ceremony. The amphitheater space connects Flinders University and the Office of Indigenous Strategy & Engagement with the campus landscape. The amphitheater frames a ceremonial space with a central fire pit and a sculptural shelter called a Wardli. Designed by a Kaurna artist and Wiradjuri architect, the Wardli provides shelter and supports the sharing of cultural practices. Building on principles of minimal disturbance and celebrating Indigenous concepts of Healing Country, the amphitheater is formed through a series of mesh decks. The decks provide all abilities access across the steep landscape and the mesh enables sunlight and rain to reach the earth below, allowing the uninterrupted re-vegetation of the site.

Essendon

This residential garden was planned alongside the home designed by FIGR Architecture Studio. The aim was to create a responsive garden which was highly sensitive to the new home. The garden design explores ways of dissolving the interface of home and garden, blurring the touch points of the building and the landscape. For the most part, the external paving does not meet the house walls deferring to garden beds located as intermediary spaces softening the architecture. In addition to this, paving borders were staggered, bleeding into adjoining garden beds with uneven edges. This was achieved by the use of full and uncut old recycled bricks laid in a running pattern. Over time the paving will age gracefully with very little effort on the client's behalf.

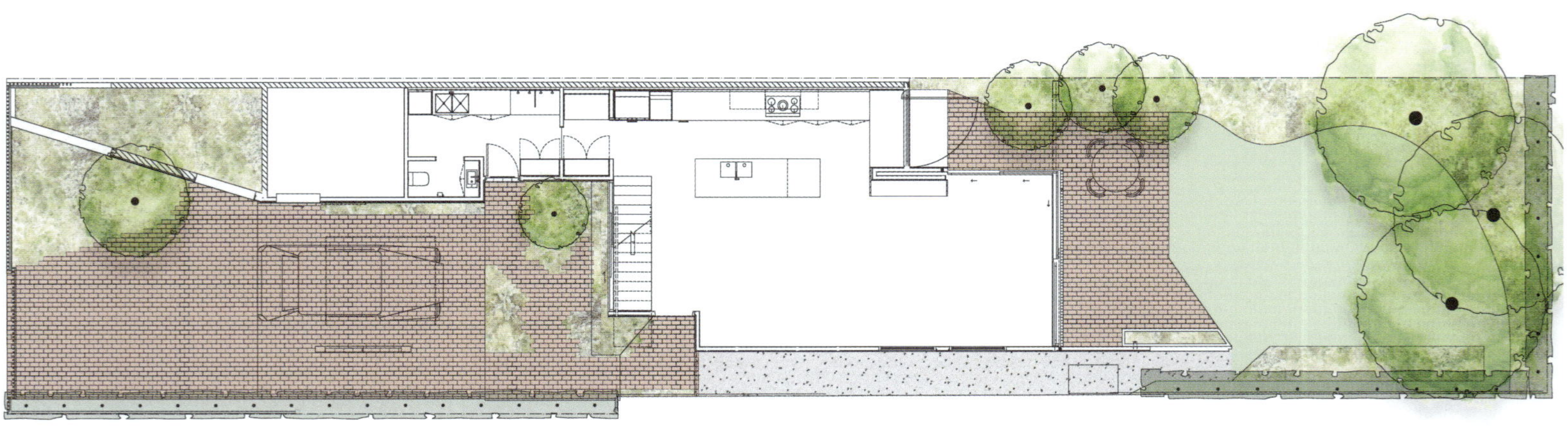

The Roundtable

The Roundtable is a temporary communal table, seven meters in diameter, filled with edible plants. The plant species have been selected for their appearance, smell, and taste. Visitors are invited to pick, smell, and taste the plants, but also to think of others as they do so. The project was installed for several months as part of city of Melbourne's post-COVID City Activation grant program. The main body of the table base was made from marine grade form ply, the surface was mild steel with a clear penetrol sealant. The seats around the table were off the shelf pots filled with sand for stability. This allowed them to be removed, emptied, and then refilled with plants from the center of the table. By activating an underutilized section of public parkland, the Roundtable revitalized the urban space and fostered a sense of community engagement.

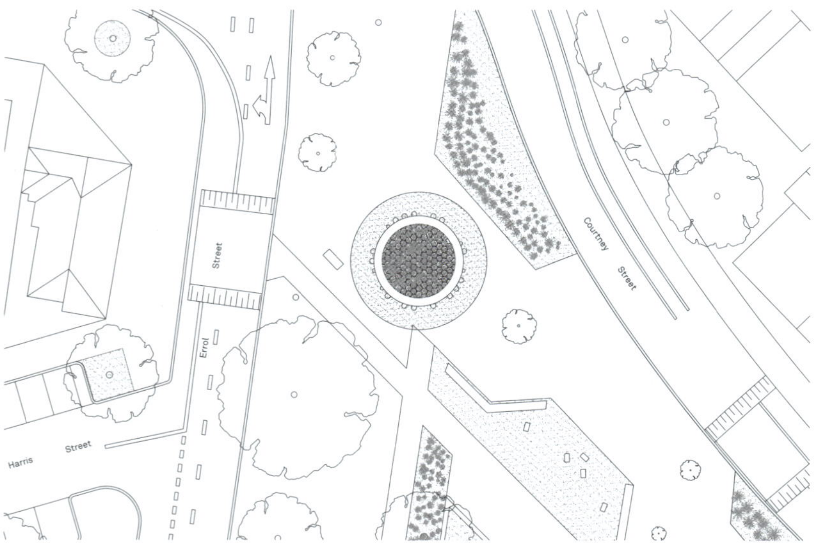

Thyme
Thymus Vulgaris

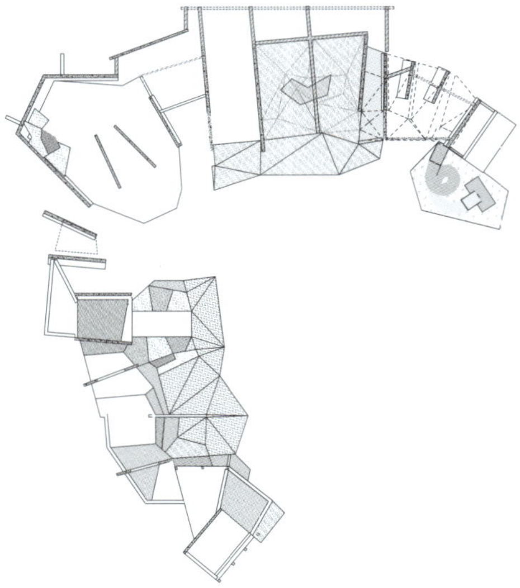

Frew Park Arena Play Structure

Built on the grounds of the former Milton tennis stadium, the Arena Play Structure honors the site's history with design references to court lines, surfaces, and tennis equipment through use of color and materiality. The complex structured design features tennis-themed play elements and street art, creating a challenging yet safe play environment for kids to discover their risk tolerance. It offers players slides, climbing walls, swings, nets, and ropes, hang-out spaces, and interactive spaces. The grandstand is brought to life with large precast concrete panels of undulating heights and angles, suspended concrete slabs, angled steel struts, steel mesh tunnels and barriers, and angular folded roof canopies arranged in an arc to enclose a central lawn space. The fragmented forms capture the energy and movement of tennis in action.

Fairbridge Childrens Park

Between 1938 and 1973 nearly 1,000 British disadvantaged children were forcibly migrated to a farm school in Molong, rural Australia. Their lives at Fairbridge Farm School were a limiting and traumatizing experience, now subject of official UK and Australian Government apologies. The site selected and park developed by Clouston for the Old Fairbridgians Association aim to be a reflective experience for visitors of all ages – positive, but conveying the poignant story of the children's journey to Australia and through life. Its design structure is evocative of this journey, evolving through precincts which in character and interpretive features express the childrens' origins, passage, farm life, and turn out. The park is not simply a place of rest, but of reflection – it represents an antidote to the hardship and isolation of its former residents.

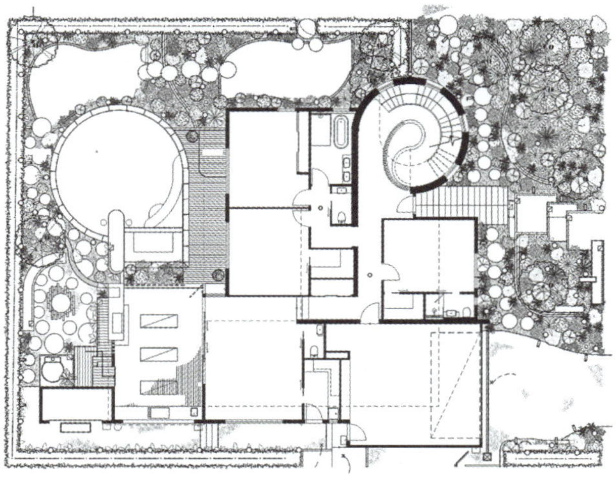

Mornington Project

Central to the brief for the backyard was the ask for a pool and spa, a covered alfresco area, fire place, pizza oven, lush coastal gardens, level lawn areas, a storage shed, and road access. The home boasts a large red brick tower to the front of the property, so a round pool, hugged by graduating limestone fingers, was created as a contrast. The pool and spa help terrace the sloping rear yard and create seamless functionality and a real sense of space in the tight area given. The juxtaposed curved flowing lines, eclectic coastal plantings of clipped spheres, lose and flowing grasses along with the privacy screening of the bay laurels hedges all work beautifully to soften the strong masculine form of the home, while the advanced topiary olive grounds the garden and connects it to the surrounding Mornington Peninsula perfectly.

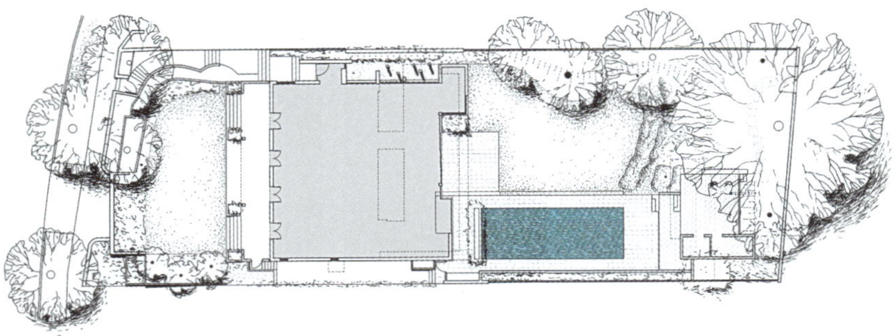

Between Predictable and Unexpected

The property has many small gardens, the first of which, in an extremely narrow courtyard to the side of the house, contains a grove of tall burnt tree trunks. The trunks look like strange creatures waiting in line, a scene frozen in time. The dead, blackened trees are quite confronting, yet small life survives within them, with tiny tillandsias clinging to the trunks, which stand in a bed of large pieces of broken slate. This narrow courtyard leads to a simple fern garden, with slate and bamboo a unifying element between the two yards. Bizarre green hills, lit up at night, standing like a miniature version of a mountain range shrouded in mist, were created with a pyramidal polystyrene base, with hessian, growing medium and baby tears (Helxine soleirolii) held in place with bird netting.

Rotorua Lakefront

Lake Rotorua is a taonga – a highly prized natural resource – for those that live on its shores, the city that has grown along its edge, and the nation that treasures it as a destination. The lakefront redevelopment re-establishes the connection between land and water through a boardwalk, inspired from a traditional Tau Kōura fishing line – anchored at one end, floating at the other and reflective of Kōuramawhitiwhiti – a gathering place referring to the movement of Kōura below the surface. The journey alternates between land and lake to break down the boundary to water, strengthened with crossing tukutuku bridges physically binding the boardwalk to the land. The design allows the public to reengage with the water and activate the lake edge while clearing away the clutter to reveal the significance of Mokoia Island.

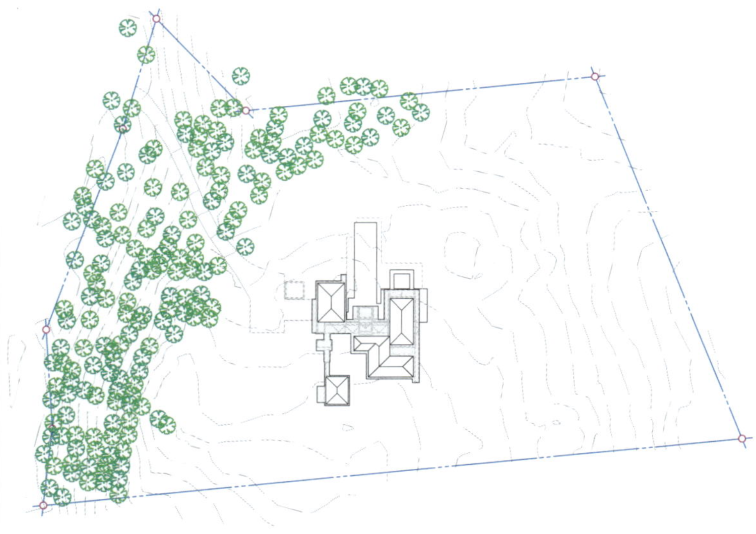

Dune House

Dune House is a holiday home located to the north of Auckland. The site, essentially a sand dune, offers magnificent views over the oceans as well as a golf course. Due to the harsh and windy conditions, the house is set low into the land, with flat timber roofs cantilevering out to form generous terraces. Stone clad pillars anchor the horizontal planes of the roofs to the ground and frame the panoramic views. A Moroccan limestone walling was used, which mimics the tone of the sand, in combination with oiled cedar for both wall and ceiling planes. Color accents are provided by rusted Corten steel and the peacock blue of the pool. The landscaping not only responds to the dune environment, but also enhances the references to mid-century American modernism in the architecture.

Peak-A-Boo

Peeking out from among the trees on the historic grounds of the 1969 Woodstock Music and Art Fair, the project takes shape as a continuous series of wood-laminate arches and decks which form a pavilion and flexible performance space. The installation stands as the first piece of programmable infrastructure in the Bindy Bazaar woods and marks the beginning of the second phase of a three-year pilot program to develop a signature art and architecture festival at the site. The project was meant to reengage the historic site with design-build pedagogies and was created by Texas Tech University students as part of the summer course led by Neal Lucas Hitch and i/thee. The result is a soaring structure, light as a feather but stiff as a board, cascading through the tree canopy like a wooden waterfall or a magic carpet.

Boston Children's Hospital Green Masterplan

MYK-D worked with the client and the design team to create a Green Masterplan for the transformation of Boston Childrens Hospital main campus in the Longwood Medical District. This project pays special attention to the growing need for acute and complex care. To more effectively connect the gardens to patients, caregivers, and families now and in the future, the design was customized to meet individual needs. This plan includes a new vertical hospital with state-of-the-art technology and the ability to admit more patients to the Boston Childrens Hospital. The patient and family garage and the main building will be connected by a pedestrian walkway, allowing more efficient and improved inpatient and outpatient care. Patients and families have already taken advantage of the Rooftop Healing Garden, the first of the gardens.

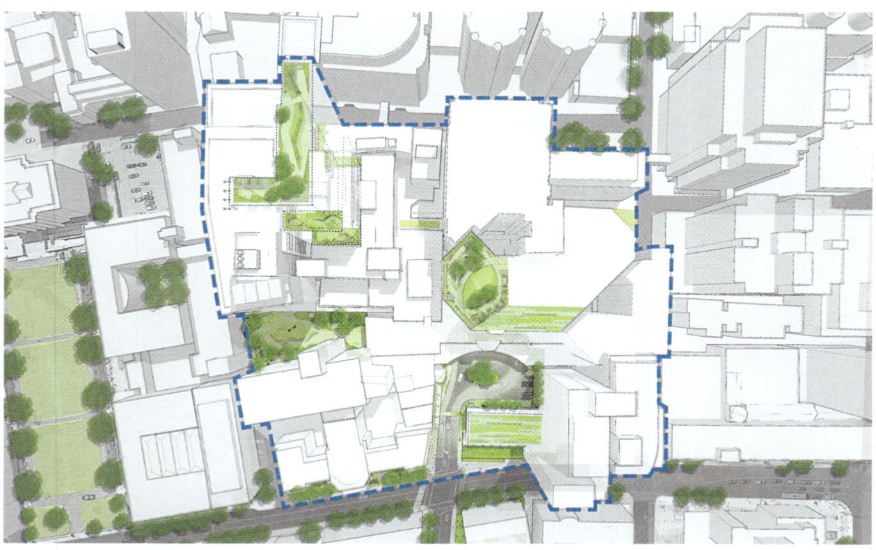

Tontine Crescent Tactical Plaza

The Tontine Crescent Tactical Plaza is located on Franklin Street, a historic crescent in Boston's Downtown Crossing. The design for Tactical Plaza is a reimagining of the 18th century design by Charles Bulfinch. He envisioned Tontine Crescent as the city's first garden plaza, where elm trees shaded a public garden. The last tree disappeared in the Great Boston Fire of 1872. The redesign of the square was the first step in reclaiming the space for its original purpose as a walkable green space. The surface of the plaza was painted with a bold plaid pattern overlaid with large graphic tree canopy symbols. The bright green makes the park vibrantly colorful. Oversized planters screen the plaza from the rest of the traffic. A protected bike lane along the edge of the plaza allows for safe bicycle travel.

Bank of America
East

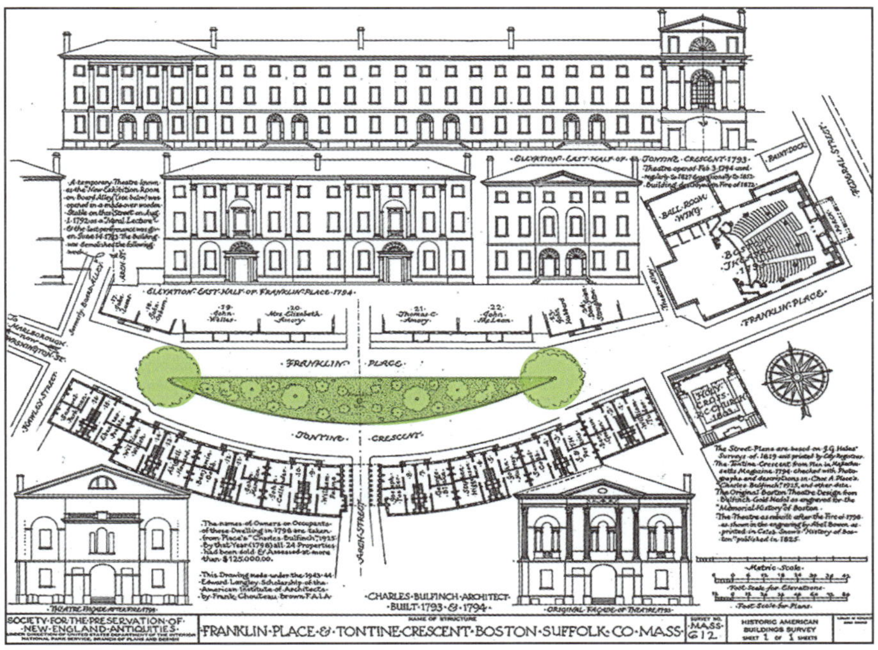
FRANKLIN PLACE
TONTINE CRESCENT
ARCH STREET
FRANKLIN PLACE
The names of Owners or Occupants of these Dwelling in 1798 are taken from Place's "Charles Bulfinch," 1925. By that Year (1798) all 24 Properties had been sold & Assessed at more than $125,000.00.
This Drawing made under the 1943-44 Edward Langley Scholarship of the American Institute of Architects by Frank Chouteau Brown F.A.I.A.
CHARLES BULFINCH ARCHITECT
BUILT 1793 & 1794
THEATRE FAÇADE AFTER FIRE 1798
ORIGINAL FAÇADE OF THEATRE 1793
Metric Scale
Foot Scale for Elevations
Foot Scale for Plans
SOCIETY FOR THE PRESERVATION OF NEW ENGLAND ANTIQUITIES
NAME OF STRUCTURE
FRANKLIN PLACE & TONTINE CRESCENT BOSTON SUFFOLK CO MASS
SURVEY NO. MASS 612
HISTORIC AMERICAN BUILDINGS SURVEY
SHEET 1 OF 1 SHEETS

David Street Station

The Downtown Development Authority (DDA) Board, comprised of volunteers, wanted to revitalize the community economically and socially through downtown growth by providing free activities in an outdoor gathering place focused on families in the area. The visual identity celebrates the place's connection to the local energy industry, the American West, and its proximity to abundant natural beauty. The plaza's grid is reinforced with string-light covered promenades that internalize circulation and provide effortless wayfinding. The promenade connects two main programmatic areas: A performance pavilion with a terraced concert lawn on one side, and a splash pad that transforms into an ice rink in the winter on the other. Their orientations play with adjacent street grids, sun angles, and wind for maximum comfort.

→ The Urban Conga → Cultural → 2022 → Chapel Hill, NC → USA → Photos: Tom Waldenberg, Anna Routh Barzin, Christopher Brickman

pArc

The pArc project is an open-ended programmable space designed to better connect the Ackland Art Museum to the surrounding community. The design both mimics and challenges the museum's Georgian-style architecture to create a softer edge to the building. It takes the symmetrical composition of the museum and breaks it down into a series of interconnected arcs that frame different social spaces. The space embraces the identity of its users and uses its open, playful design to break down social barriers and create a sense of community. It responds to the user and the environment by manipulating light both during the day and at night. Shadow play elements evoke moments of joy and wonder as the sun sets. Thus, pArc becomes a flexible communal landmark that creates endless ways to play, gather, perform, teach, or converse.

→ Skidmore, Owings & Merrill, Urban Rivers → Public Parks and Gardens → 2021 → Chicago, IL → USA → Photos: SOM | Dave Burk

Wild Mile

The Wild Mile creates a new environment for habitat, education, and recreation on the Chicago River. Designed as a 169,800-square-meter floating eco-park, it advances a community-led vision of renewed urban ecology that helps strengthen neighborhood connectivity, clean polluted waters, and support a healthy, robust river ecosystem in the heart of the city. The development for the City of Chicago, Department of Planning and Development, transforms the formerly industrialized canal into a one-of-a-kind park that serves people and wildlife, It provides walkways, kayak docks, and other amenities, restoring the river as a public trust. The eco-park is the result of collaboration between Skidmore, Owings & Merrill, Urban Rivers, and Omni Ecosystems, with support from Tetra Tech, D'Escoto, O&H Community Partners, and Near North Community Partners.

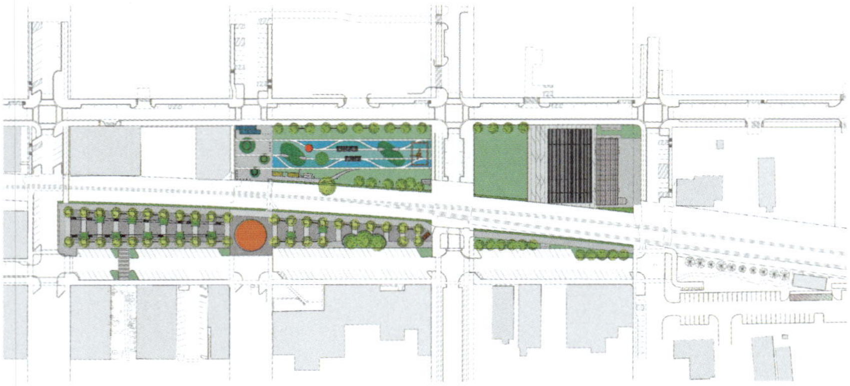

Railyard Park

With a grant from the Walton Family Foundation, the City of Rogers established a new downtown park to make this historical part of Rogers a regional destination. A citizens survey collected over 1,000 responses and helped Ross Barney Architects to define the needs of five priorities: Inviting, Memorable, Challenging, Beautiful, and Authentic. Rogers first plat of survey, completed in 1881, uses the rail to create a strong dividing line. The park's design ignores the rail as a barrier and stitches the east and west together. The result is a series of plazas that can transform throughout the day, week, month, and year. These versatile and flexible spaces create a new and distinct rhythm that extends beyond the park into adjacent streets. They turn this piece of downtown to a new center instead of an edge.

NG YOU FL
WATER
STOP

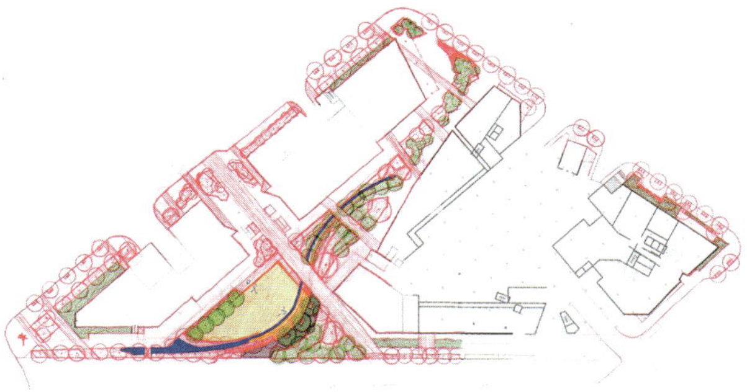

Ivy Station

Ivy Station, located adjacent to the Culver City Metro E line station, provides a welcoming and active gateway for residents, metro riders, and visitors. The 20,000-square-meter public park runs north to south through the development, connecting the Metro station platform to the Venice/National Boulevard intersection. It is programmed with a host of activities, such as games, fitness, movie nights, art shows, gift fairs, and concerts. The park's planting utilizes a significant collection of California native plants and plant communities. The plan evolves and transforms from high elevation, dry planting into a riparian landscape. The landscape is underscored by an urban nature level that integrates reclaimed storm water runoff. The project also performs beneficial storm water cleansing, connecting the site to the city's urban fabric.

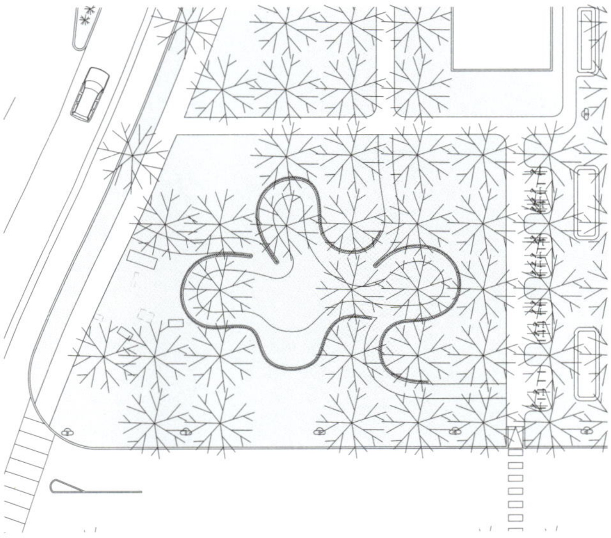

Mirage

Located in the olive grove adjacent to the Visitor's Center at Apple Park, Mirage includes over four hundred glass columns that weave through and amongst the olive trees. The pillars of pure cast glass are made with sand sustainably collected from desert regions across the Earth, flowing like a dune shaped by the wind. In daylight, the sculpture varies in iridescence, and in the evening it gently glows. The experience of watching the light fall through becomes a form of meditation. Visitors can interact with the artwork, walking alongside and through, where the glass appears to subtly melt into the landscape, like a desert mirage. Mirage is a global artwork, celebrating each of the lands from which it is created, and the people who nurture, conserve, and sustain these lands.

Tarpon Bend

The project Tarpon Bend shows that the design of this home underscores the firm's ongoing explorations of regional modernism by adapting a rectilinear frame with site-specific and climate-specific considerations. In this case, vertical fins were integrated into the rectilinear frame addressing shading and privacy concerns. This house allows the practical responses to the site and climate to become architectural identity of the buildings. The project takes full advantage of its unique location at the confluence of the Tarpon and New Rivers near downtown Fort Lauderdale. Aside from the pleasant riverside setting, the property features a grove of gnarly oak trees. The footprint of the home was designed to accommodate these characterful and elevated old specimens and integrate them perfectly into the design.

1173

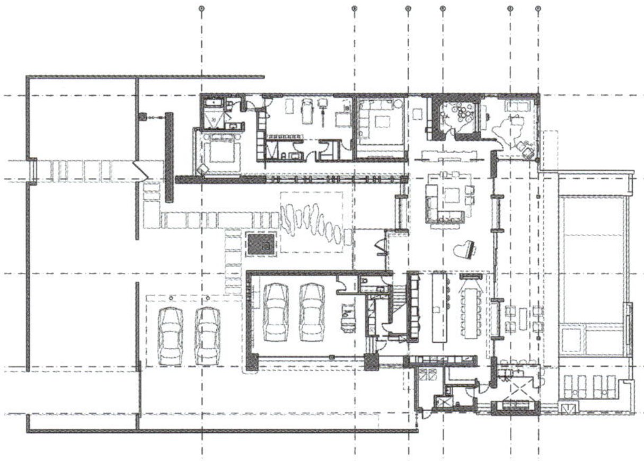

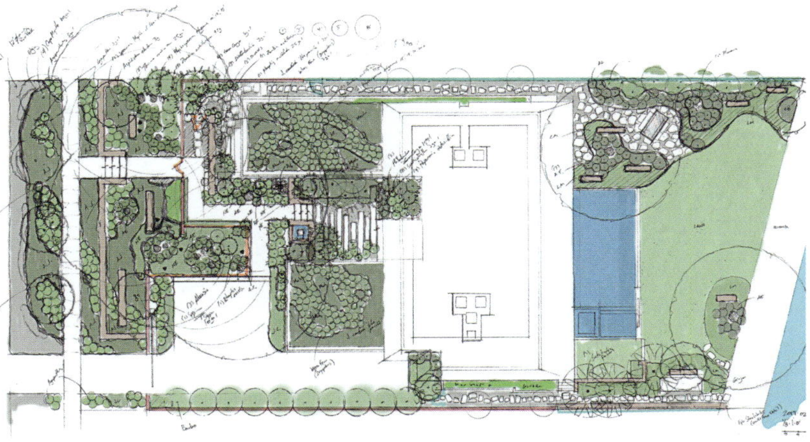

+85
+69
MEANDERING OLIVE GROVE
+53
BELVEDERES W/ PARAPET HEIGHT WALLS
SPECIMEN OAKS
+37
+29
+21
OHLONE ACADEMIC
M+S

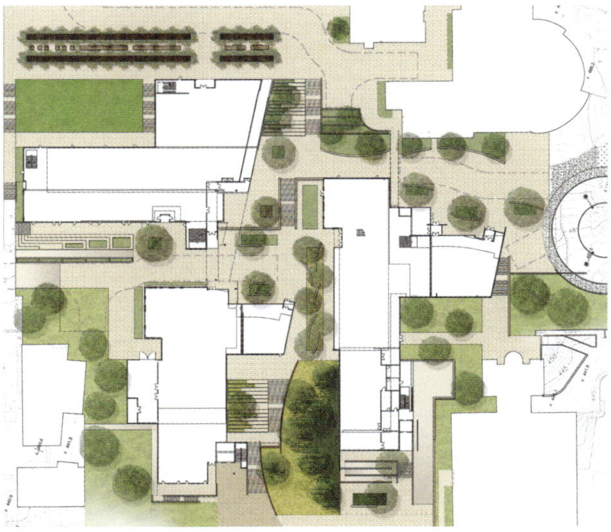

Academic Core Ohlone College

The new science building, arts center, and library were composed to naturally terrace into the sloping foothills. Connected by grand staircases, terraces, and belvederes, the new academic core takes advantage of views to the western horizon and provides social gathering spaces dedicated to the various academic surroundings. The selected landscape materials are simple, sustainable, and native. A stylized meadow visually connects the heart of the campus to the rolling foothills to the east. Coastal live oaks weave through the meadow and creep over paved terraces, providing shade, and a gathering place at the entrances to each building. Staircases are embedded in the surrounding landscape and offer greened steps for an engaging experience.

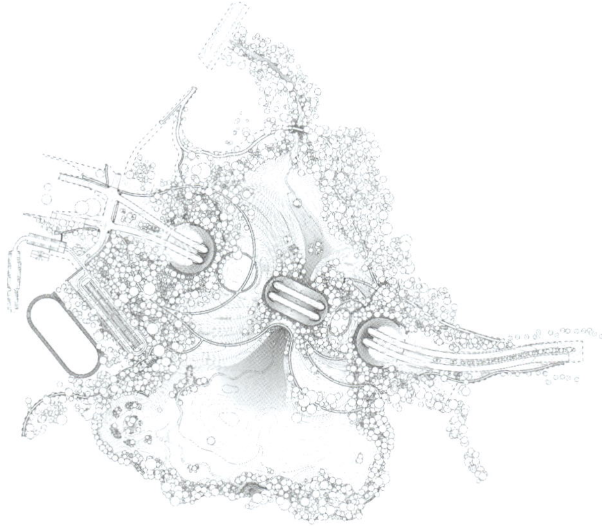

Kinder Land Bridge and Cyvia and Melvyn Wolff Prairie at Memorial Park

The Kinder Land Bridge and Cyvia and Melvyn Wolff Prairie is found at the heart of Memorial Park, spanning over Memorial Drive. The project restored around 180,000 square meters of native coastal prairie and savannah ecologies, realizing a multi-use trail network and creating an expansive moment where park users can cross and enjoy Houston's urban skyline. Prairie restoration adjacent to the Land Bridge reintroduced endangered native Gulf Coast prairie species and expanded areas to include additional wetlands. Both prairie and wetlands serve important habitat and stormwater management functions, which is an essential part of the park's ecological restoration. With a diverse mix of grasses and forbs, the expanded prairie complex provides year-round visual interest and important habitat for nesting birds, reptiles, and insects.

License to Chill

Once featured in the James Bond film License to Kill, the garden has been significantly updated and now offers stunning views from every angle, especially from the elegantly designed wraparound porch. Designed to emphasize the interplay of tropical plantings, a myriad of contrasts in size, shape, color, and texture draw the eye throughout the garden. The design plan focused on a variety of interconnected spaces ranging from an outdoor kitchen and dining area to a pool, and a secluded spa with an outdoor chandelier. The garden was filled with a variety of tropical plants that vary in leaf size, shape and color, creating a lush oasis with as much native planting as possible. Most importantly, the garden needed to be a place you never wanted to leave.

Laguna Canyon Foundation

After the floods of 2014, which washed up waste from centuries of farming, three terraces were created as part of the clean-up, which served as the starting point for 20,000 square meters of landscaping. They were connected with staggered timber stairs, and a large swale, to soak up and redirect waters that may flood the canyon, was dug and planted. The client, Laguna Canyon Foundation, and Laguna Beach community wanted to keep the touch light, and foster a collaborative spirit in order to make a functional and beautiful working and dwelling space for the volunteers, employees, and nature who share this public-facing open space. Stumps were added on the discovery terrace, and timber seats at the gathering terrace. On the utility terrace up top three open-air nursery structures enable seedlings for future restoration work.

PRISCILLA'S GARDEN

Angel Oaks Garden

Angel Oaks Garden is dedicated to its pre-existing trees: Quercus virginiana (live oaks). A landscape paradigm shift of gardening as a work of anamnesis, ecological restoration of site and the re-generating of bio-diverse native plant communities. No-lawn-no-concrete, permeable planting beds, diverse planting perimeters open to pollinator gardens within the subtropical hammock, supporting the diversity of pollinators and sustaining rich food webs. The garden flows into the architecture with cascading Ernodea littoralis and Jacquemontia pentanthos over-hanging pollinator gardens. They transform place and architecture to live-habitat gardens, alongside the companion species.

Times Square

On average, more than 300,000 people use this public space in the heart of New York City every day: tourists, onlookers, commuters, customers, strollers and road users. The redesign by Snøhetta, carried out over eight years, unifies the appearance of Times Square and Broadway from 42nd to 47th streets. In addition, the width of the sidewalks on 7th Avenue was significantly increased, resulting in new urban space here too. The routes for pedestrians were standardized, made safe, and adapted to the different speeds of the various groups of users. Among other things, different pavements and street furniture – including granite benches more than 15 meters long – were used for this purpose. The stairs on the TKTS Booth and the redevelopment of Father Duffy Square were done by Perkins Eastman as architects.

Pier 26 at Hudson River Park

Located in Hudson River Park near the Tribeca neighborhood, Pier 26 stretches out and over the Hudson River with striking views. The 10,100-square-meter pier is located within the Hudson River Estuary, a highly sensitive and ecologically productive body of water, which contains a mix of freshwater runoff and ocean saltwater. The pier has a unique mission. It provides a physical and virtual space to experience the unseen dynamics of the Hudson River, reflecting current technologies and scientific understanding of its health, ecological successes, and challenges. The physical design of the park is a dynamic gradient experience from upland to lowland, from land to water. This experience provides ecological education and recreation for people of all ages.

Hunter's Point South Park

Hunter's Point South was envisioned as an international model of urban waterfront ecology and a world laboratory for innovative sustainable thinking. What was once a barren post-industrial site has been transformed into a world-class park that is both urbane and otherworldly. Employing innovative sustainable design strategies, the team has woven landscape, infrastructure, and architecture into new public open spaces that resonate with the surrounding communities and connect neighbors and visitors to each other, the riverain heritage, and spectacular views of the Manhattan skyline. Its core of water taxi, café pavilion, multi-purpose lawn, and active recreation is balanced with grassland promontories, an iconic overlook, and a native habitat marsh whose storm resiliency doubles as a compelling urban waterfront trail.

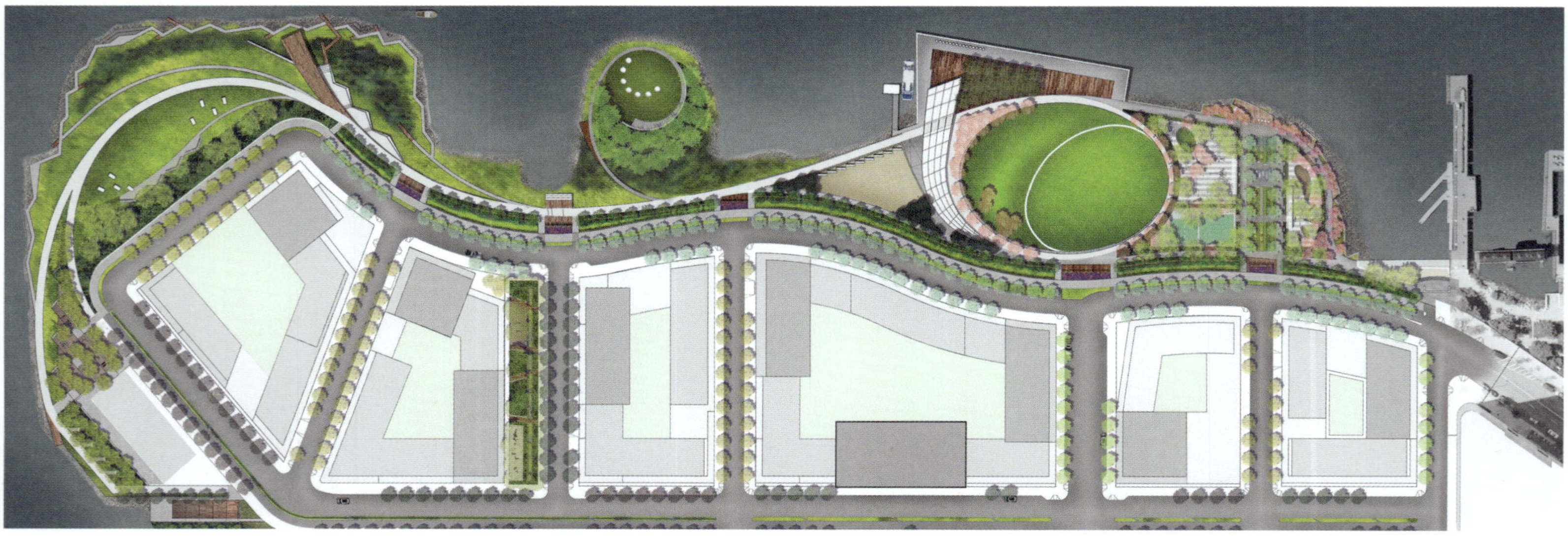

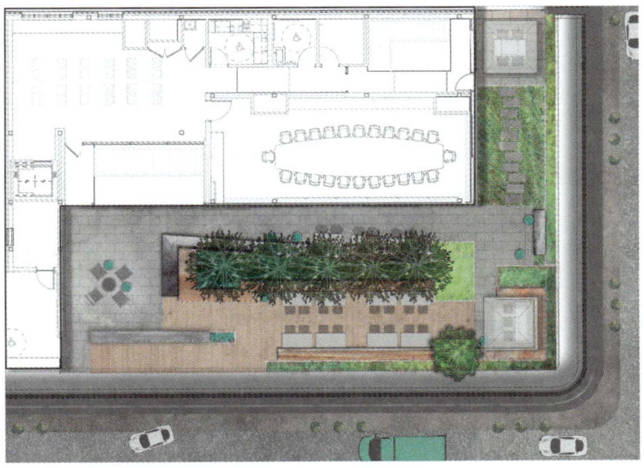

The Flatiron Institute

Named after the architectural focal point of the historic neighborhood, the Flatiron Institute is dedicated to thought leadership in the fields of mathematics, astrophysics, and biochemistry. The garden expands the Center's world-class programs into a vibrant landscape, positioning nature as an essential partner in interdisciplinary discussions. Creative professionals can be found engaging in interdisciplinary dialogue on the tiered amphitheater seating. Multi-layered elevations of benches and stadium-style seating encourage unexpected gatherings. The outdoor learning rooms feature chalkboards and areas for stargazing. Greenroof plantings combat the urban heat island effect, while absorbing storm water. The Flatiron Institute's rooftop garden and interior plantings thoroughly integrate nature and the office environment.

Little Island at Pier 55

Little Island at Pier 55 is located on the Hudson in the west of Manhattan. It replaces Pier 54, where the survivors of the Titanic sinking once landed and which was damaged by Hurricane Sandy in 2012. Unlike the other piers, the square island picks up on Manhattan's street grid. Two equally rectangular bridges connect Little Island to the mainland. It consists of 132 tulip-shaped concrete elements ranging in height from 4.6 to 18.9 meters, which give the construction an undulating appearance. The park design comprises 35 different tree species and is characterized by very different areas of use: There are lawns, a secret garden, fitness areas, scenic observation areas, plazas, and performance spaces.

Pier 57

Located at the epicenter of New York City's prominent Hudson River Park, adjacent to Manhattan's popular Meatpacking District, historic Pier 57 is an iconic and storied structure dating back to 1952. The studio !melk was commissioned to design all of the pier's exterior spaces. On the roof of Pier 57 is New York City's largest rooftop park with picnic areas, observation decks and a restaurant. The rooftop park offers spectacular panoramic views of the city and the Hudson River. The park is open to the public throughout the year. It is home to many programmed activities and cultural events. Transformed from its original maritime industrial use through an award-winning historic restoration, Pier 57 is undoubtedly one of New York City's most memorable waterfront experiences.

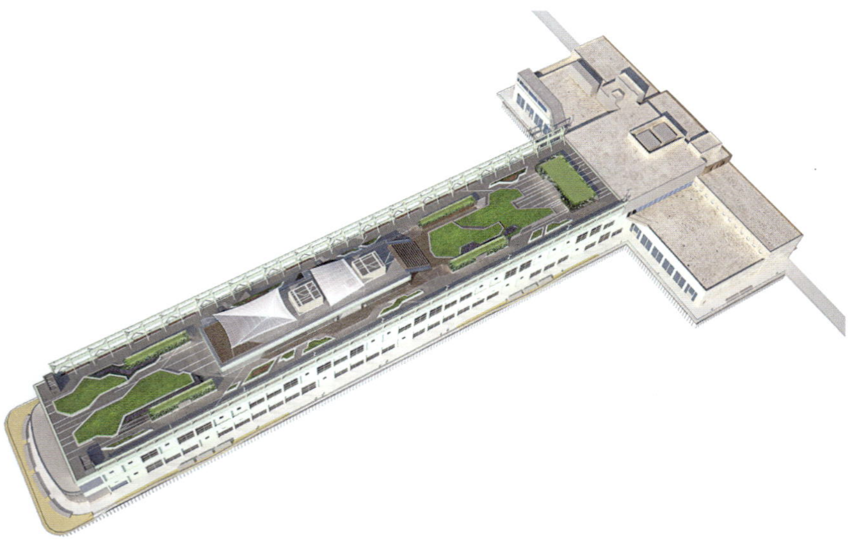

Ingadi

Ingadi is a historic farm in upstate New York at the confluence of two rivers. Three historic barns overlooked abandoned farmland that had lost all character and was severely impacted by invasive species. The project team – Beverley Kerzner/The Phantom Gardener/Mancini enterprises – has ecologically restored the wetland and woodland areas and enhanced several zones by creating softly sculpted wild meadows that form paths and views to architectural and natural focal points. Extensive re-grading of worn-out paddocks has resulted in a sculpted lawn slope leading to a new farm bridge. Naturalistic stone work, the sinking of the new driveway into the ground, careful offsetting of meadows from fences, and similar large-scale sculptural moves inscribe the man-made landscape into the boundaries of forest and river.

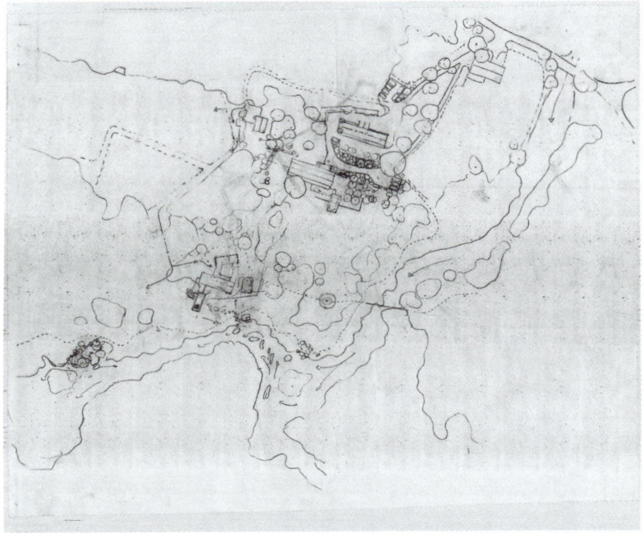

The RiverFront

The RiverFront is a compelling example of how urban planning and design can revitalize riverfront cities. Comprising 29,000 square meters of downtown parkland along the Missouri River, the project scope includes the transformation of three underutilized parks into a connected and activated public realm: Gene Leahy Mall, Heartland of America Park, and Lewis and Clark Landing. Improving community mobility and access between the three designated parks is a key feature of the design. The goal is to make the river more accessible to the community while providing inviting outdoor spaces and fostering important connections within the city. The trio of parks provides improved access and connectivity within downtown Omaha, promotes environmental connectivity, and supports economic growth.

→ Floor Associates → Private Gardens → 2016 → Paradise Valley, AZ → USA → Photos: Christopher Brown, Bill Timmerman

Nita Residence

Designed to blur the lines between indoor and outdoor spaces, this private residence celebrates the dramatic landscape and climate of the Sonoran Desert. The modern, transparent home is organized around a series of courtyards and gardens that serve as extended living spaces. These distinctive courtyards draw residents from within the glass interior to an exterior composition of color, texture and pattern in the immersive desert landscape. Each courtyard is unique and includes features such as an outdoor shower in the master bathroom, a Zen-like water feature at the main entry, integrated cast-in-place concrete seat walls, and a linear fire feature. Specimen desert trees, including ironwoods and palo verdes, provide filtered shade for the owners' dramatic collection of succulents and cacti.

Kokedama Forest

Kokedama Forest is a 100-square-meter miniature sculptural forest consisting of more than 1,200 autochthonous forestry seedling kokedamas nested within a 2.5-meter-high inverted dome carved out from a fragile and intricate metal network woven with steel, inviting the contradictory blend of strength and delicacy that is implicit in all living networks. Kokedama Forest is a tribute to the humble dwellers within the soil and how their partnership with plants creates a suitable habitat for all of us. Soil is a habitat that is normally overlooked; however, these micro-organisms are indispensable as they are actively building fertile soil while nourishing their key partnership with plants through an intricate network of relationships that constitutes the core of healthy and resilient ecosystems.

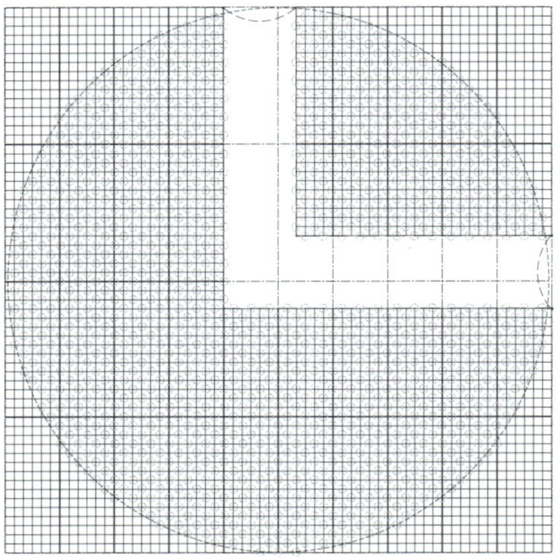

Leach Botanical Garden

The redevelopment of Leach Botanical Garden by Land Morphology modernized a formerly languishing garden, enhancing ties between the site and program to ensure future financial stability and growth. The project gives visitors many compelling reasons to visit the natural environment at the 173,000-square-meter garden owned by Portland Parks and Recreation in Oregon. Land Morphology first produced a long-term plan with public input, owner and team collaboration. A tree walk elevated almost eleven meters over a forested slope, plus a pollinator meadow, a gathering green, and a pavilion with fireside terrace were realized. The project developed the upper garden, establishing a new system of trails and stairs. An adjacent 5,000-square-meter pollinator meadow with 240 unique taxa attracts insects and birds.

Roger Williams Park Gateway Center

Located on Broad Street, next to the historic Roger Williams Park, the Roger Williams Gateway Center, planned by InForm Studio Architects is intended to improve community access to the park, inviting visitors to celebrate this incredible public asset. Following various transit, bike, and pedestrian routes, there are many cultural, educational, and retail destinations, activities, and institutions leading from downtown into the park. It introduces amenities and services to encourage and expand the rich cultural identities. As a Net Zero Energy project, it produces its own energy through on-site photovoltaic panels and is economically sustainable in the long run. The landscape retains all water on-site and is a microcosm of a garden called People's Pleasure Ground, leading visitors through a transition of spaces from urban to wild.

ROGER WILLIAMS PARK
GATEWAY CENTER

The Urban Green

The Urban Green is a captivating 8,100-square-meter public park that celebrates its natural surroundings. Inspired by the majestic Sierra Nevada Mountains and the scenic Truckee River, the park reflects the beauty and spirit of its locale in Reno, Nevada. The park's bespoke design elements showcase a tailored approach to creating spaces that are functional, visually striking, and distinctive. With emphasis on a hierarchy of circulation paths, movement, and exploration are encouraged. The park offers a variety of experiences, from vibrant shared spaces to more secluded areas, catering to a diverse range of visitors and activities. Environmental art elements integrated throughout the park stimulate the senses and foster, creativity and interaction. Seating is placed to capture sightlines of the western sunset and frame distant mountain peaks.

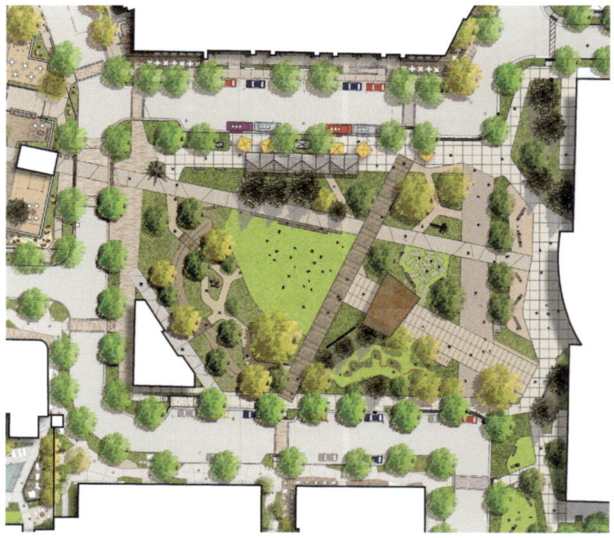

Piney Branch Water Garden

The Piney Branch Water Garden creates a connection and transition between the built realm and its ecological context. The project is located at the Biomedical Sciences and Engineering Building at the Universities at Shady Grove in Rockville, Maryland. Harvested rainwater emerges from a raised source at the end of the garden, flowing through a curved series of shallow pools and copper spillways. Three sculpted Light Columns emerge from the garden at the far end of the courtyard. The sculpted aluminum details throughout the garden are abstractions based on natural patterns, observed phenomena, systems, and structures that reference the scientific and engineering disciplines found within the building. The project includes a green wall with sculpted aluminum details just inside the building.

→ Deneen Powell Atelier → Education and Playgrounds → 2022 → San Diego, CA → USA → Photos: Deneen Powell Atelier

Wildlife Explorers Basecamp

This immersive, interactive children's zoo includes indoor and outdoor exhibits for mammals, reptiles, and insects. Interspersed throughout the animal exhibits are loads of opportunities for simultaneous visitor parallel play – climbing, running, clamoring over bridges, splashing through waterfalls, and jump jets – mimicking animal behaviors. There are four distinct environments – desert, tropical, marsh, and woodland. This 12,140-sqaure-meter tapestry of diverse habitats transports visitors to a land that is rich in adventure with a multitude of opportunities to explore nature. Deneen Powell Atelier also designed the artistic and immersive donor elements and animal and insect graphics throughout the site.

Lattice House

Carefully folded into a western hillside overlooking Richardson Bay, this coastal home is situated on the edge of a precipice, maximizing full panoramic views from San Francisco to Mount Tam while still maintaining intimacy among the live oaks and Monterey pines that envelope the site. The project's formal language was inspired by terraced landscapes and shade canopy structures that respond to the steeply sloping terrain, while protecting the occupants from exposure to the harsh elements of the surrounding natural world. The resulting architecture creates a home that is both grounded and dynamic, reaching out to the site's spectacular views while instilling the occupant with a sense of protection and comfort.

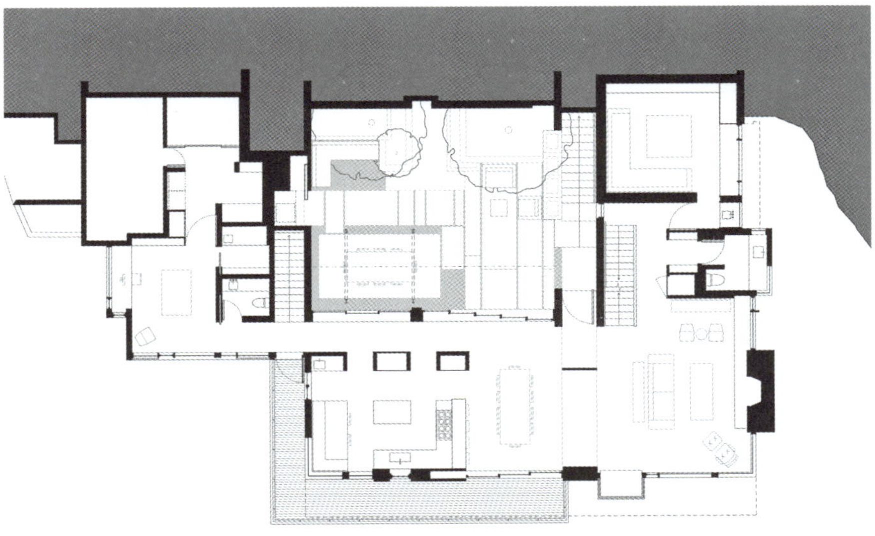

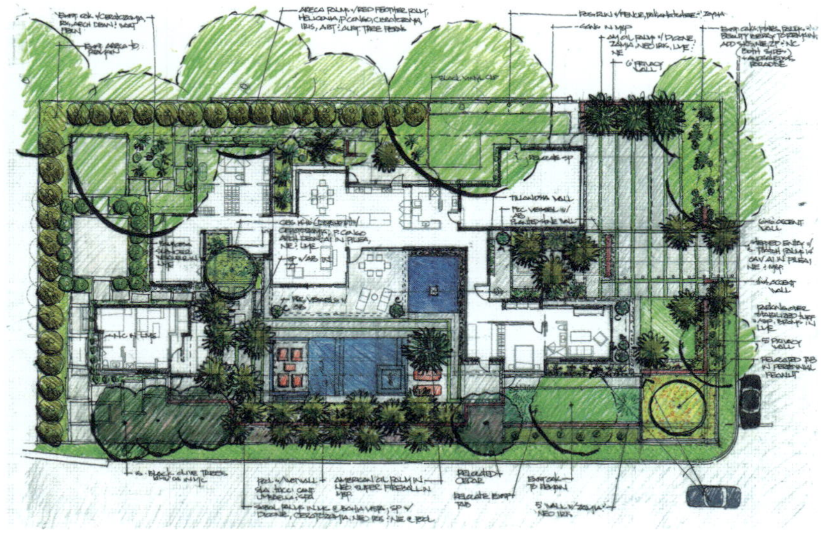

Citrus Avenue

With its use of open floor plans and large panes of glass, the residence manifests a clear geometry floating above the Florida landscape, allowing the homeowners to experience the landscape from each interior space, providing uninterrupted connection to the environment and daylight. The site facilitates the homeowners' active lifestyles with a lap pool, a meditation garden, and clear pedestrian circulation that lets the owners and neighbors interact in a meaningful way. At the heart of the site is the courtyard, where a covered porch steps down into the pool garden. The long southern exposure of the courtyard is densely planted to mask sight lines from large multistory houses across the street, and a vibrant Barragan-esque accent wall with a scupper spillway circulates water to muffle the road noise.

Expedia Group HQ

The Expedia HQ waterfront site provides an innovative approach to the workplace with a focus on sustainability. Soil remediation and stormwater management were key drivers in the site's design, resulting in a systems-based approach that centered on long-term resilience, while also celebrating the plants and regional landscape of the Pacific Northwest. Planting around the site is salmon safe and contributes to habitat formation both within the site and across the larger waterfront. There are nine distinct planting areas, each with its own character. The planting palettes pay special attention to soil restoration to support a more holistic systems-based approach to landscape maintenance centered on land stewardship. The project's public component, called The Beach, celebrates the land-water threshold.

Bending Arc

Bending Arc is a monumental sculpture composed of 1,662,528 knots and 290 kilometers of twine. It gently billows above the Pier District, allowing the wind to create a choreography of ever-changing shapes in the sculpture's soft surface. The sculpture's color also changes moment to moment as its surface interacts with natural and projected light. During the day, the sculpture casts shadows on the park and people below, and at night it transforms into a glowing beacon of magenta and violet light. The aerial sculpture spans 130 meters and measures 22 meters at its highest point. The artist titled the sculpture Bending Arc in reference to Martin Luther King's words: "The arc of the moral universe is long, but it bends toward justice."

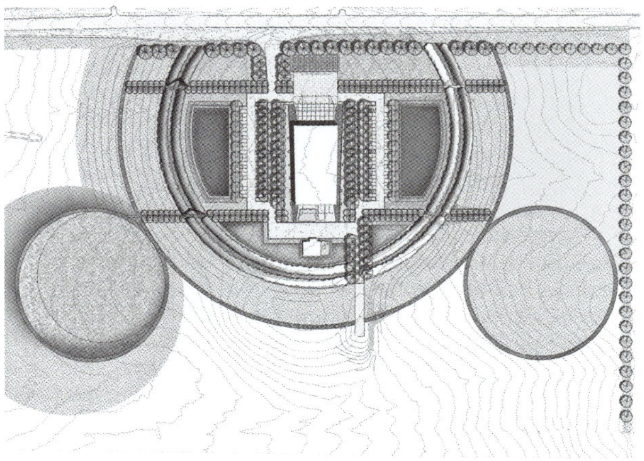

Pritzker Museum, Library, and Archive

The Pritzker Museum, Library, and Archives endeavor to create a non-profit research institution designed to enhance the public's understanding of military history and the sacrifices made to maintain freedom and democracy. The grounds and architecture are evocative of the simple geometric shapes of historic military earthworks. The site is laid out around a circular rampart, a reference to a military defense strategy, with the museum and associated program elements at its center. The circular site is divided by linear pathways that provide unique spaces for wetland gardens and restored prairie meadows. Allées of oak trees flank the pathways, providing a human scale while simultaneously spanning the length and connecting the visitor to the wider site.

Prairie Line Trail (PLT)

A depository of memories and experiences, Prairie Line Trail corridor (PLT) is a reinvented landscape affectionately called the High Line of the West. Its storied history dates to 1873 when the Tacoma port became the western terminus of the Northern Pacific Railway, connecting the Great Lakes to Puget Sound. Decades later, as the railroad began ridding itself of the 1980s wide right-of-way, the University of Washington re-purposed the industrial infrastructure into public open space unifying downtown Tacoma, Brewery, University, and Museum Districts with the waterfront. Serving as a steward of watershed environment and catalyst for green development, a beloved Prairie Line Trail celebrates the regions' ecological and civic ethos.

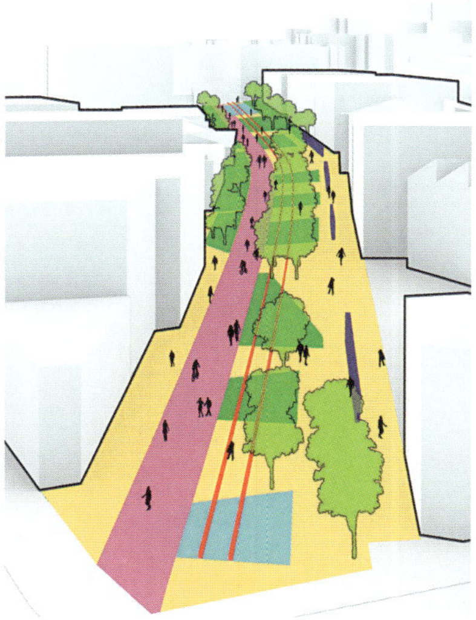

Vancouver Landing at Terminal 1

A vibrant gateway to Vancouver's redeveloping waterfront breathes new life into a derelict amphitheater. The Vancouver Landing project is both a holistic expression of Vancouver's reinvigorated connections to the Columbia River and a model for landscape architecture-led adaptive reuse. As the project's prime consultant and world's first salmon-safe-accredited landscape architecture firm, GreenWorks, P.C. led over ten consulting firms to develop commercial and residential opportunities, public green spaces and restored habitats, and a public market building in the midst of Vancouver's waterfront renaissance. Vancouver Landing links the city's waterfront, downtown, and trail systems. This project, with Vancouver Landing as the first phase, rebalances Vancouver's ongoing relationship to the Columbia River.

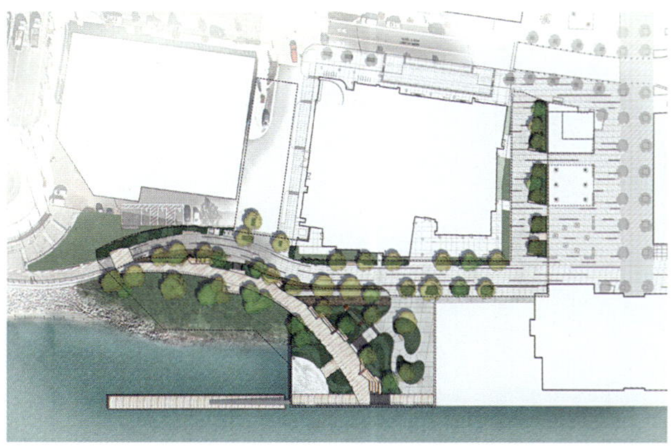

→ Civitas, W Architecture and Landscape → Public Parks and Gardens → 2015 → Calgary → Canada → Photos: Angus Media, CMLC

St. Patrick's Island

The bow island became a site of homelessness, drugs, and sex crimes after the former campground was closed. Calgary Municipal Land Corporation planned a new bike and pedestrian bridge linking East Village with the island and the light rail station just north. They reached out to the public to create a master program plan for the island to be restored as a premier example of biophilic design. Civitas and W Architecture won a design competition to create an active, nature-based park on the island for residents of the growing downtown and nearby populations. Analysis revealed that the island had two breaches in the past, which were destroyed by later development. The design restored a functional breach and wetland, and used those excavations to create a mound that provides a prospect upriver and to downtown.

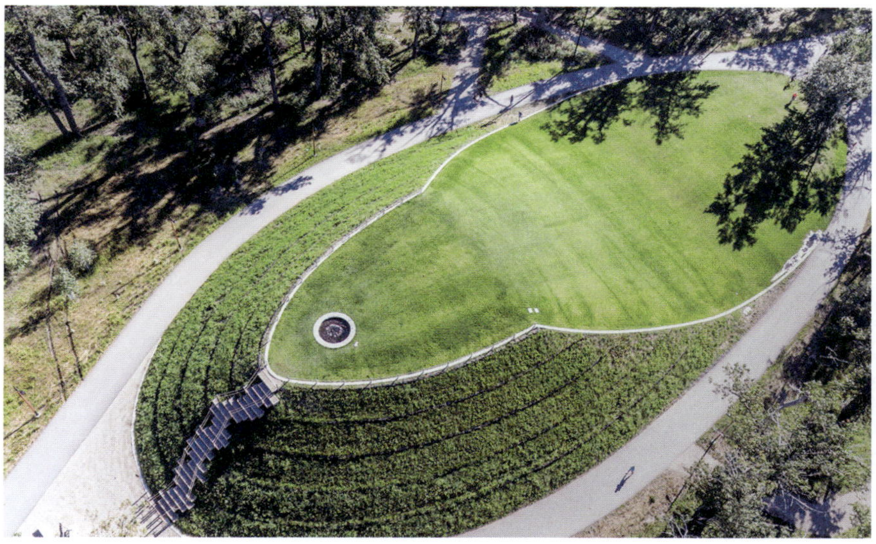

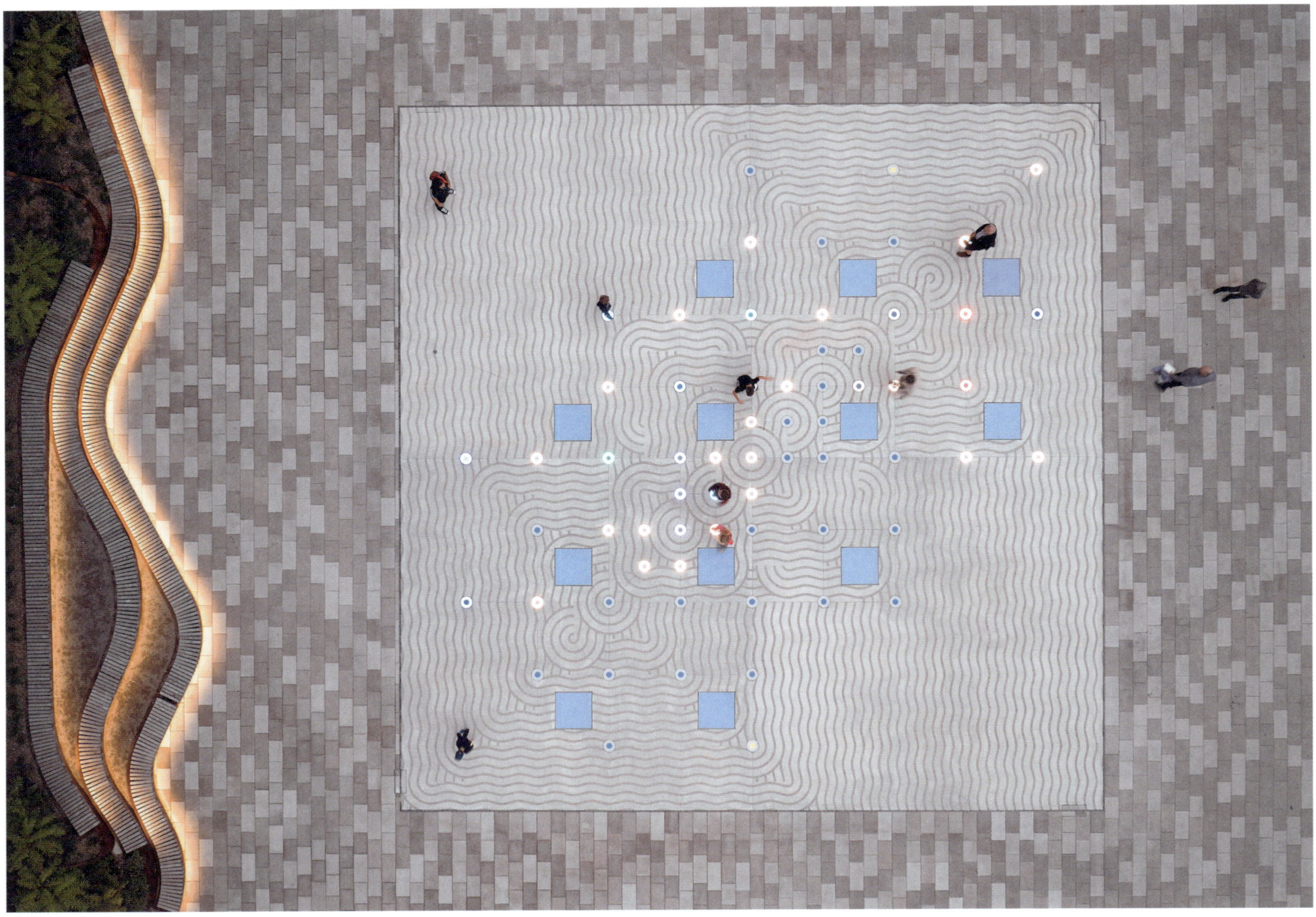

River Lines

River Lines is one of two Daily tous les jours pieces commissioned for the Gaslight District by HIP Developments, as part of their vision for "Joy Experiments"—interventions designed to foster community through play. The interactive, wave-patterned pavement embedded with 62 sensing light rings gets people moving through musical collaboration exercises. Developed through research, interviews and engagement workshops with the local community and developer, the artwork is a response to the historical nature of the site. It is an example of how investing in the public realm is essential to reimagine and activate places and encourage collective interactions. River Lines shifts the city's attention back to its long-neglected waterfront, its design, and musical score.

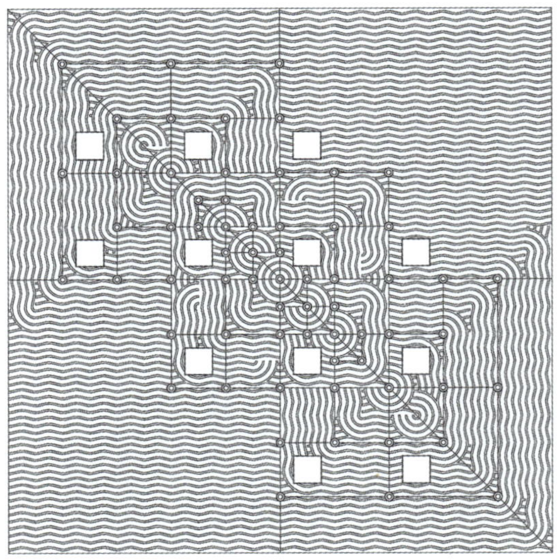

Carvolth Integrated Open Space / Goldenview Park

Carvolth Integrated Open Space is a 16,200-square-meter linear green space located in Langley's Carvolth neighbourhood at the headwaters of the Serpentine River watershed. The Space was designed to receive stormwater from the surrounding 63,000-square-meter mixed-use site through a series of visible features that slow, clean and celebrate the water before it enters Latimer Creek. With water as the central thread, the concept thoughtfully combines form and function, weaving a curated collection of functions, programs, and communities into one public space. Water droplets delineate spaces and programs while also indicating the direction of water flow through the site, while visible stormwater features reinforce our understanding of and connection to the systems active on the site and to the larger ecosystem.

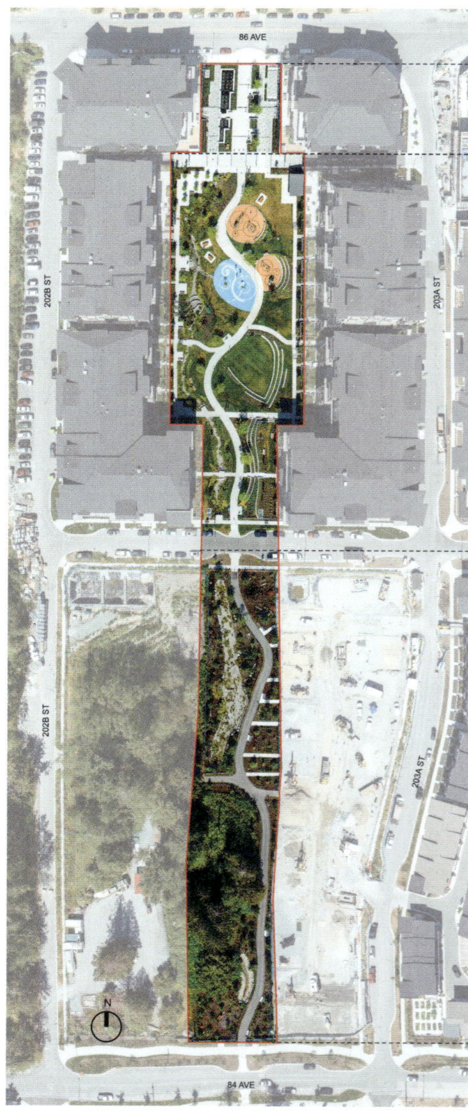

Grand Quay of the Port of Montréal

The Grand Quay project, a collaboration with Provencher_Roy Architects, surpasses the commission for a world-class ferry terminal by offering a large public space and providing access to the St. Lawrence River, while highlighting a remarkable historic ensemble that bears witness to the creation and economic development of Montreal. The fruit of close collaboration between architects, engineers, and landscape architects, it revitalizes the heritage landscape and, in an exemplary fashion, rebuilds the links between the city and the river. The project dares the colossal maneuver of lowering the quay structure to provide access at water level. Every simple gesture in this project is in dialogue with the grand scale, highlighting its fluvial and industrial identity.

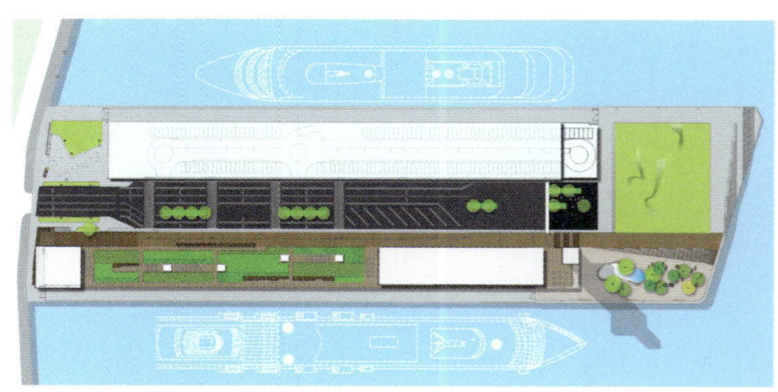

Franciscan Commemorative Park

This small commemorative park, located in Montreal, Canada, stands on a site formerly occupied by a neo-Gothic chapel, erected by the Franciscan Order at the end of the 19th century. The Borough of Ville-Marie decided to use the empty site as public space for the more than 105,000 residents in this district. Inspired by the building's rib-vaulted ceiling and its delicate rosettes, the landscape designers generated a unique pattern of diagonals and points. Planted beds were added to this unusual composition, subtly blurring the historic reference. The only artefact recovered from a fire that destroyed the chapel in 2010 is the bronze bell now guarding the park entrance. It acts as a subtle counterpoint to the formal and material abstraction of this urban oasis where memories of the past are kept alive.

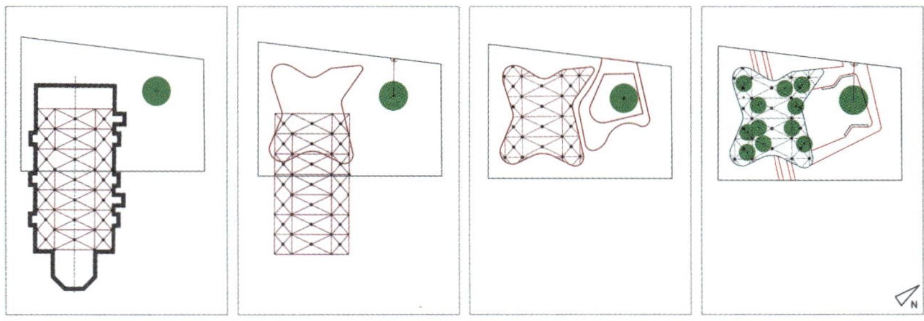

Agora Maximus

Agora Maximus is a three-year tactical urbanism project rethinking ubiquitous temporary installations. As such, it proposes a modular, sustainable, and green solution. Located in the Quartier Latin, it celebrates its legacy as the founding site of Montreal's Francophone institutions. Agora Maximus channels two archetypes of Latin culture by merging the Agora, a place for civic gatherings, with the Circus Maximus, an esplanade for live entertainment. Its iconic figures feature an array of pixelated mounds in assertive colors: gray evokes heritage, white embodies modernity, and orange dares a nod to road construction signage. A minimalist palette of indigenous plants greens each urban oasis. The installations are linked together by a striking pattern of orange-hued pixels on the pavement to enliven the streetscape.

Votre gouvernement
Québec

Théâtre de Verdure

The Théâtre de Verdure, an emblematic site in La Fontaine Park, has hosted countless shows since its inauguration in 1956 until its temporary closure in 2014. To breathe new life into this enchanting site, the Lemay team developed a concept based on the relationship between architecture and landscape, theater and park, the built and the natural. This reflection is part of a broad, inclusive, and pluralistic vision in which the development relies on continuity with the landscape to provide a coherent and relevant support for the architectural intervention. The architecture of the Théâtre de Verdure frames the landscape of La Fontaine Park by opening up to the park, dematerializing its façades, and maximizing views. The new venue thus functions as a theater, a park, an object, a garden, and an open window on the magnificent site.

Place du Sable-Gris

Place Sable-Gris reinforces the cultural and artistic vocation of this section of Ottawa Street. The name refers to the early metal casting technique developed at the Darling Foundry, known as Sable-Gris (gray sand). The design of the plaza reflects the genius loci of the site by celebrating the neighborhood's industrial past in a contemporary way. The large square, unified by a textured surface, is inspired by castings seen under a microscope. The event portal evokes the memory of the gantry cranes that characterized the now defunct industrial activities of the area. This contemporary artifact adds a third dimension to the public space. Finally, the soft landscape expresses the process of transforming urban and industrial environments into more natural spaces: highlighting the poetry of vegetation reclaiming the asphalt.

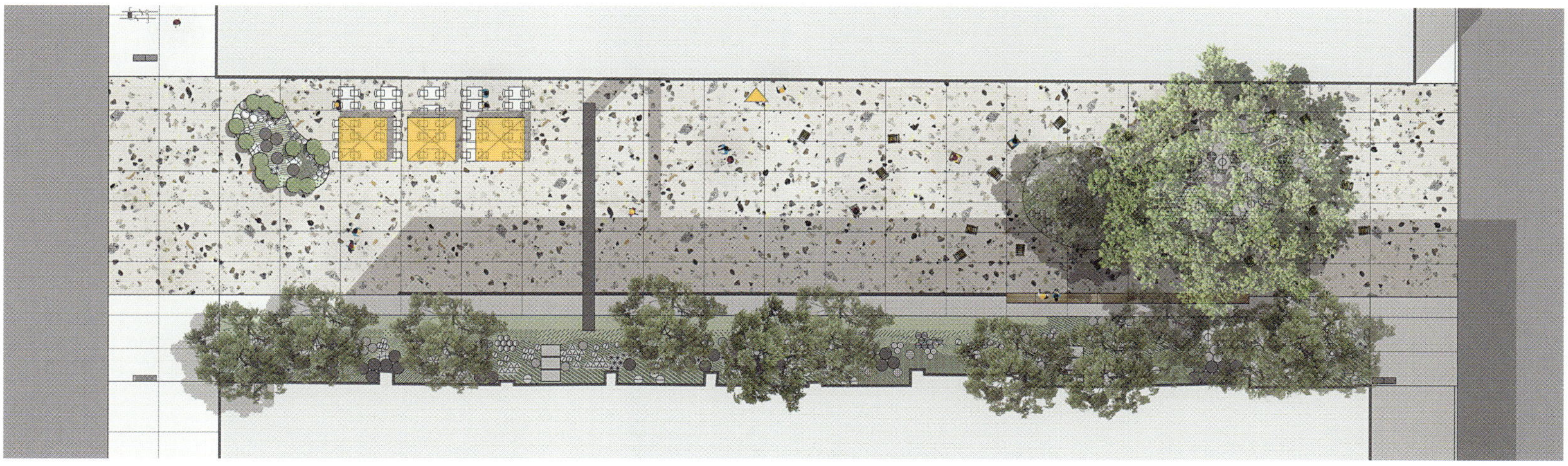

Berczy Park

Berczy Park is an iconic little park positioned at the threshold between three Toronto neighborhoods – St. Lawrence, the Old Town, and the Financial District. The gradient from historic façades to a backdrop of skyscrapers creates an unforgettable setting for this resolutely urban park. Its redesign created a plaza with a large fountain and detailed granite paving motif, a green space for passive activities, and a perennial garden next to the historic Flatiron Building with its iconic 1980 mural. Pathways running diagonally through the space establish long trajectories and sight lines to make this small park feel big. The increase in the number of dog owners has been manifested into a signature and playful fountain that merges a classic Victorian two-tiered fountain with a menagerie of water-spouting life-sized dog figures.

Trilium Park and Wiliam G. Davis Trail

Formerly an employee parking lot, Trillium Park examines the relationship between culture and nature while symbolizing the geological and vegetative evolution of Ontario. Landforms such as drumlins and a sculpturally abstracted escarpment, reminiscent of the effects of a glacial moraine, provide the topographical structure for the park. The site's peaks and valleys create the illusion of additional space and privacy, while the winding network of trails that encircle the site lead visitors to moments of discovery that range from grand vistas to engaging park features. One such feature is the granite ravine, a gateway experience that celebrates Indigenous placemaking through the Moccasin Identifier Project. The native plant palette has transformed the site into a sustainable botanical wonderland.

Manulife Head Office Exterior Renovations

This exciting open-air office project began with the vision of a campus-style environment focusing on flexibility, wellness, and the surrounding natural landscape. A terrace links directly to the cafeteria, providing a large outdoor dining area. From the terrace, a serpentine and accessible pathway seamlessly winds through the tree canopy as it gently slopes down to connect to a larger trail system. The elevated boardwalk ingeniously links a sequence of well-placed open-air office pods. Through this design, the site has been transformed from an uninspired horticultural landscape typical of office settings, into a naturalized landscape that attunes the employee's work experience to the rich and unique setting of the surrounding ecosystem and reinforces the client's sustainability commitment.

Cosoleacaque Main Plaza

The project aimed to improve the quality of social life, considering basic design strategies such as flexibility of use, reuse, and optimization of existing elements in good condition, and the sustainable use of materials with high resilience that are locally available. Due to the climatic conditions, like the subtropical-humid, the ancient cultural roots from the pre-Spanish era and the tight budget for the project, the entire public space, the surrounding walkways included, was fundamentally redesigned. A key strategy is the provision of shade, both through architectural elements and vegetation, as well as the introduction of a new visual character inspired by elements of local culture, for example colored horizontal stripes, circular shapes, reinforcing people's sense of belonging in this special and atmospheric place.

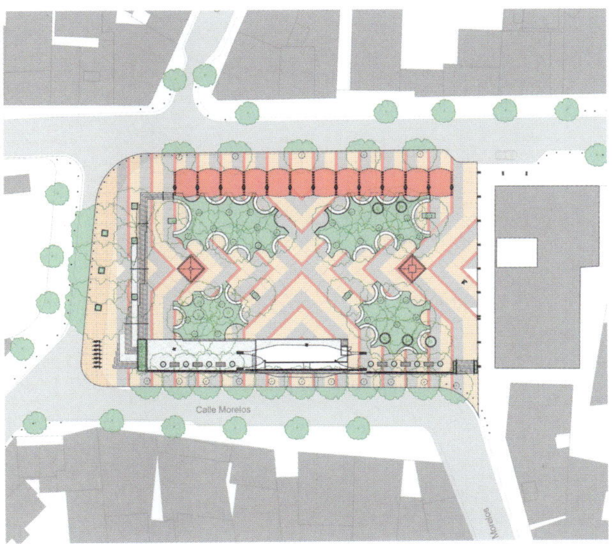

CIVAC Linear Park

CIVAC Linear Park is an infrastructure and urban renewal project of one of the main avenues of Jiutepec. It activates a series of public and cultural spaces in the Industrial City of the Cuernavaca Valley (CIVAC). The project integrates a linear park, a civic center and a new skate park. The Linear Park is located on a 28-meter-wide and 1.15-kilometer-long boulevard between two busy roads. The design consists of a continuous line that deviates in different sections to respect and highlight the existing vegetation. The Linear Park ends with a multipurpose circular forum, a circular platform with Texcal stone steps that can be used as a playground or an amphitheater. The Civic Center was designed with sand-colored pigmented concrete as the main material. It incorporates a large number of the existing endemic trees into the design.

El Terreno

El Terreno is a community garden situated on a hill in Mexico City, in the midst of a vast array of flowers, aromatic plants, and vegetables. A pavilion constructed from recycled materials utilized in the studio's previous works stands out. The project's objective was to establish a social space for the community, particularly for the youth, while simultaneously offering them an educational opportunity. The walls of the building are constructed of stones and bent iron rods, while the roof is composed of wooden trusses. According to the architects, the primary functional features of the community garden are plurality and versatility, rather than a strictly defined purpose. This allows the multifaceted space and the pavilion to cultivate new ideas and be defined by its users.

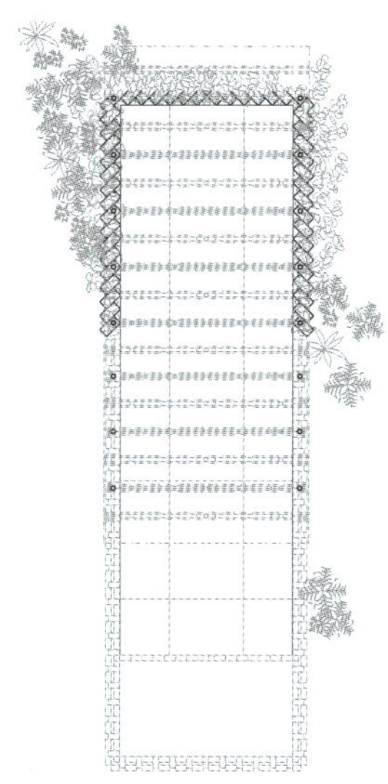

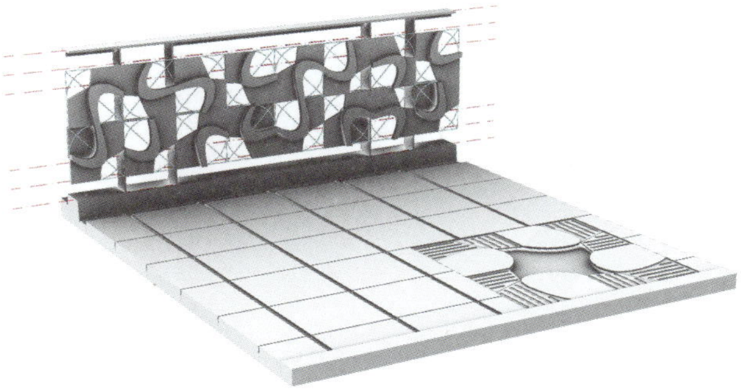

San Jacinto Square's Urban Renewal

Located in the heart of San Angel, Mexico City, for centuries San Jacinto Square has been one of the most relevant commercial and social spaces in the area. Recently, architect Alejandro de la Vega Zulueta and his firm Estereotomia developed and implemented a renovation and rejuvenation project for the public square. The project was commissioned by the INAH [National Institute of Anthropology and History], Mexico's top authority in the investigation and preservation of cultural heritage. The only remaining original element of the plaza was its layout, that allowed the architect to create an entirely new code that can be applied to other public spaces in Mexico City as well. The design of new metal guardrails to protect the square's lush green areas used parametric techniques to create new patterns.

La Carmela Cultural Center

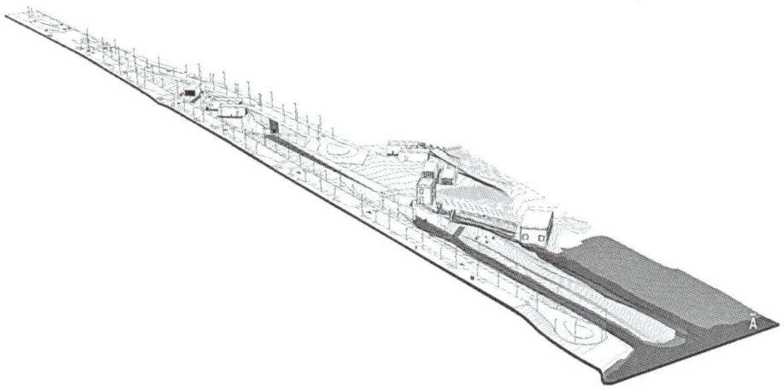

The site of the former Carmela Hydroelectric Plant, which dates back to 1906 and was abandoned 1956, has been redeveloped as cultural center and open green spaces for this neighborhood in development. Additional buildings were integrated and an urban environment with green and blue infrastructure has been established by recognizing existing natural runoff and flood-prone areas. Elements such as the pipe support blocks, tunnels, or stairs were reused as part of the hardscape for the new pedestrian circulation. The road area was reduced to offer a larger, tree-lined, and illuminated pedestrian walkway, and a linear park with recreational and resting spaces, as well as a skatepark on the enclosed canal.

El Capitan Creek Park

The purpose of the project was to create a park in the urban area of Monterrey that includes public recreational, environmental and water spaces. The park runs along the watercourse of the El Capitan stream and forms a lush and stabilized linear tree cover of 1.5 kilometers. Special attention has been given to the redesign and renaturalization of the pre-existing concrete canal, which has been transformed into a stream with trees and aquatic plants, and with a redesigned gabion system that allows greater water flow, flood control, and colonization of aquatic fauna and birds. New jogging and bicycle tracks connect to a wide existing system on the main avenues of San Pedro. The project also includes the design of playgrounds, special places for picnics, two modules with public toilets, kiosks, and viewing terraces.

Trajectory Arch Bridges

Jens J. Meyer creates works of art with great impact from little material, resulting in immaterial sculptures, by intendedly using light and space. This sculpture visualizes the trajectory of a flying golf ball thus forming the arc of two bridges. Each flying golf ball initiates a bridge and connects the take-off and landing points through the ephemeral trajectory. Just as an airplane draws arches into the sky, a bridge is created for a certain number of people, connecting the shores of islands and continents. This makes the Trajectory Arch Bridges a symbol for all bridges, connecting not only the two sides of a river, but also people, countries, and cultures. The Trajectory Arch Bridges were created in 2022 for the arte FITS Foundation's fourth Art in Golf Triennial and is part of the Ephemeral Open Air Museum at Dorado Beach Resort.

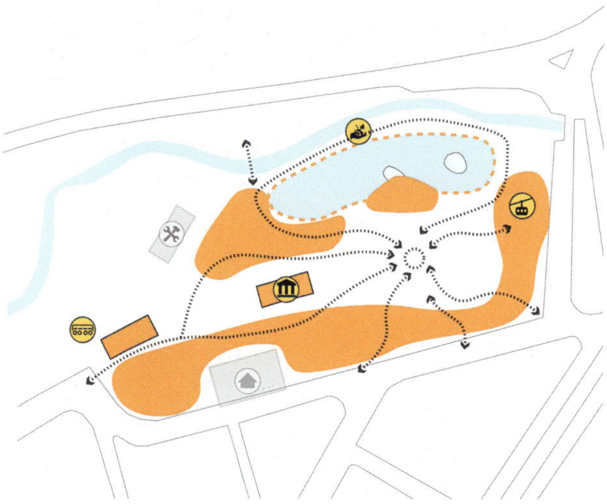

Capivari State Park

Capivari Park is situated within the communal hub of Campos do Jordão, a renowned tourist destination in the São Paulo countryside, located two hours from the city of São Paulo, which attracts four million visitors annually. Its prominence lies in its natural environment, cultural festivals, and local products. Historically redeveloped from a railroad infrastructure site, the park focuses on attracting a diverse demographic through the installation of new attractions. The local economy is stimulated through the restoration of infrastructure, services, and commercial offerings. The design incorporates existing historic structures, such as the renovated former train station that now serves as a museum, as well as natural elements. The latter was manifested through an outdoor stage by the lake and a cable car station at the top of the hills.

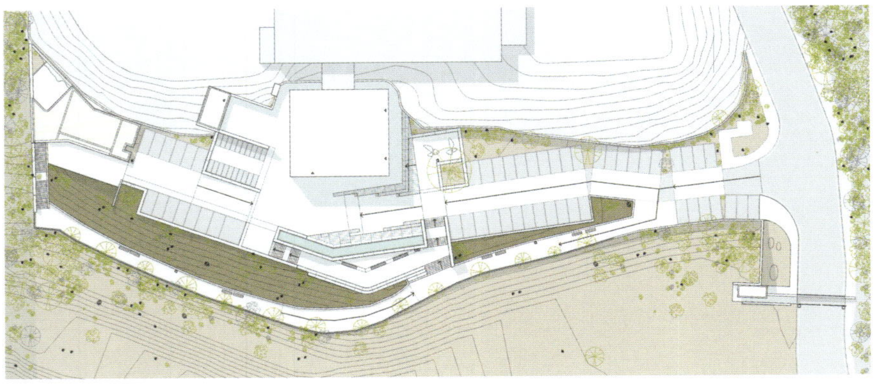

TKA Esplanade

The TKA Esplanade is a project aimed at structuring the external areas of the industrial plant of a crane manufacturer in the Serra Gaúcha region. The scope includes the design of access to the site and common areas, including the development of a plaza adjacent to the administration building, access ramps, parking, technical areas, landscaping, visual identity, and signage. It's a project where basalt rock, concrete, and metal blend in composition with vegetation and the local landscape. Each element has been carefully considered to highlight the natural beauty of the materials and the native context of the project. The light design was done by Atelier de Iluminação.

alameda das oliveiras

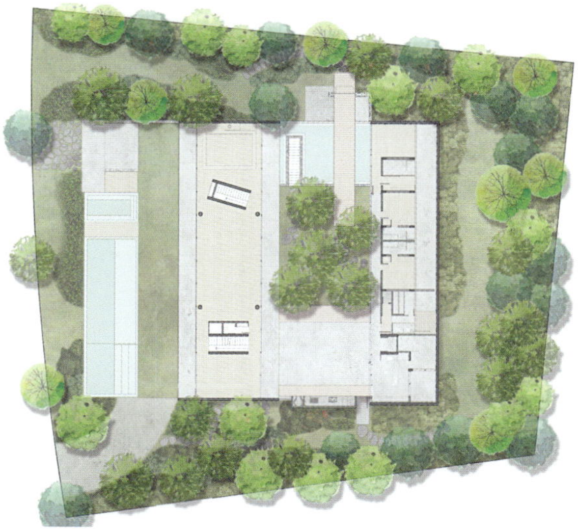

Serra Residence

Located in the state of São Paulo, Serra House exemplifies harmony between architecture and nature. It is a large cantilevered house to be shared by two families designed by architect Couto. The terrain's pronounced unevenness became an integral part of the project, resulting in an organic and lush garden that frames the architecture and deconstructs its straight lines. The central courtyard stands out in the project, separating the social area from the intimate area, creating privacy yet connection. The tropical vegetation forms an endless green backdrop, enhancing the best views of the terrain. Serra House not only offers a warm and inviting environment for its residents, but also inspires a deeper connection with nature on 2,700 square meters.

Praça da Árvore

The project idea for Praça da Árvore, which is a tree square, was developed in the community, including the children and their surrounding network. The information that was gathered provided the material for the implementation. The project management and construction work were done by the Agency for Innovation and Strategy in Recife. The aim of the project was to make more efficient use of the main element – the 12-meter-high Paineira tree. The project includes a consolidation of the floor plan, resulting in the reuse of existing floor areas and the salvaging of natural systems. This makes the project less complex and more cost-effective. On this basis, the dry fountain and the playful topography could be realized. The space was designed primarily for children up to the age of six to allow them to play freely in a green setting.

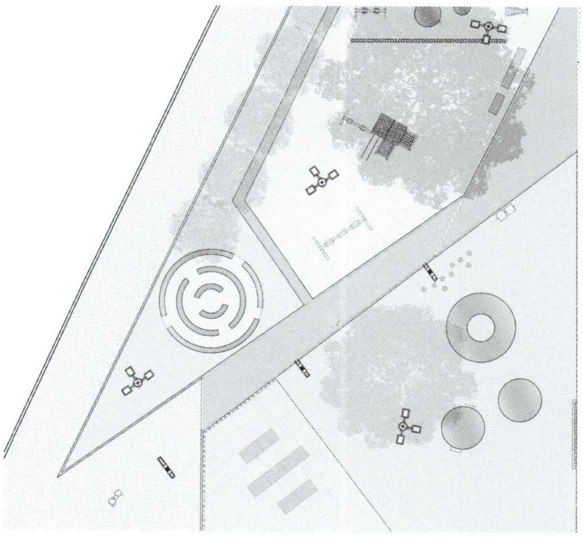

3M House Garden

A large terrace and the living room face this garden in a residential neighborhood of São Paulo, which also welcomes the social entrance to the house, the outdoor living areas, and the swimming pool, all in one place. The garden has been designed to create the magic of being discovered little by little, and to create this sensation, the vegetation, composed of trees, shrubs, palms, and foliage of different heights, shapes and colors, creates openings and closures where it is desired to protect or uncover. Graphite lavastone paves the path that weaves through this mass of vegetation to reach the pool, which can only be glimpsed from the house. The pool design, conceived as part of the landscaping project, has a living area in the same stone as the path, and the gray slate chosen to finish the pool reflects the color of the sky in the water.

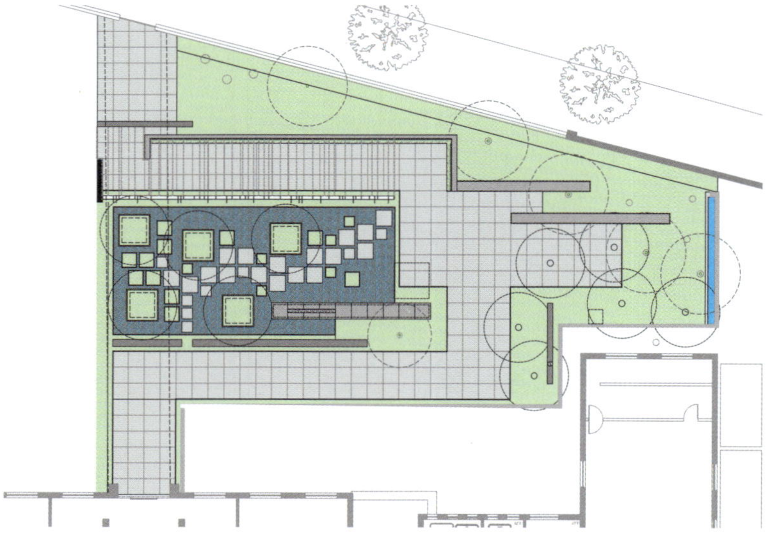

Eliane Garden

The Eliane Garden, located in São Paulo, is a contemporary lush greenery with bold, clean constructed elements, in sync with Brazilian tropical heritage. Brazil is the only country named after a tree, so the project properly honors the tree by planting trees and other native plants in harmony with the surrounding existing canopy. With a subtle incidence of natural light, the garden gives a mysterious atmosphere. At the center of the garden is an extensive reflecting pool, layered with black pebbles that were made from ceramic waste, specially developed for this project. Floating steps invite the visitor to give a closer look to the garden islands. A resting spot with benches layered with porcelain tiles provides a soothing atmosphere for relaxation. The floor porcelain tiles have an auto cleaning technology used for its low maintenance.

→ Studio MK27, Isabel Duprat Arquitetura Paisagística → Private Gardens → 2019 → Trancoso → Brazil → Photos: Fernando Guerra

Sand House

This residence by Studio MK27 reduces the functional program to the bare minimum and the confined spaces to five separate, slightly detached volumes to take full advantage of nature on the 6,570-square-meter plot. The entire structure is raised on a wooden deck and covered by a rustic eucalyptus pergola supported by 14 laminated wood frames. It filters sunlight and rain and reduces the gap between architecture and nature. The linear volume of deck and pergola is embedded in a strip of greenery designed by landscape designer Isabel Duprat. Subtly curved elements – pool, paths, plant beds – form the transition from garden to nature and coastline. The pool is aligned longitudinally at 45° in relation to the main structure, opening a dynamic dialogue between pool, house and neighboring beach.

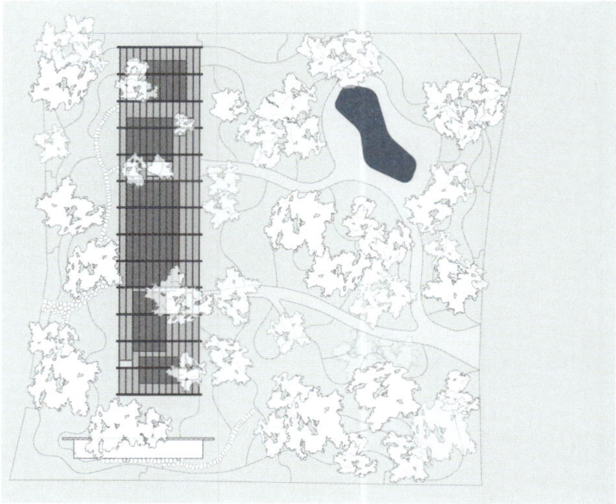

INDEX ARCHITECTS

INDEX ARCHITECTS

IMPRINT

The Deutsche Bibliothek lists this publication in the Deutsche Nationalbibliographie; detailed bibliographical information can be found on the Internet at http://dnb.ddb.de

ISBN 978-3-03768-295-1

1st edition 2024

Project coordination: Editorial office van Uffelen
Text editing and layout:
Marius Liebhart, Alisia Romeo, Charlotte Schmekel, Valeria Valuyskiy, Chris van Uffelen
Graphic concept: Michaela Prinz

All of the information in this volume has been compiled to the best of the editors' knowledge. It is based on the information provided to the publisher by the architects' and designers' offices and excludes any liability. The publisher assumes no responsibility for its accuracy or completeness as well as copyright discrepancies and refers to the specified sources (architects' and designers' offices). All rights to the photographs are property of the photographer (please refer to the picture). All plans originate from the respective designers.